Electronic Document Management Systems

Other McGraw-Hill Books of Interest

Avedon
Electronic Imaging Systems

Baker
Networking the Enterprise

Bock and Marca
Designing Groupware

Dewire
Text Management

Dewire
Application Development for Distributed Environments

Fatah
Electronic Mail Systems: A Network Manager's Guide

Hertzoff
AT&T Global Messaging: The AT&T Easylink Services Sourcebook

McClain
The OLTP Handbook

Minoli
Imaging in Corporate Environments

Minoli
Analyzing Outsourcing

Pecar et al.
The McGraw-Hill Telecommunications Factbook

Reynolds and Araya
Building Multimedia Performance Support Systems

Sokol
From EDI to Electronic Commerce

Electronic Document Management Systems

A Portable Consultant

Thomas M. Koulopoulos

Carl Frappaolo

McGraw-Hill, Inc.

New York San Francisco Washington, D.C. Auckland Bogotá
Caracas Lisbon London Madrid Mexico City Milan
Montreal New Delhi San Juan Singapore
Sydney Tokyo Toronto

Library of Congress Cataloging-in-Publication Data

Koulopoulos, Thomas M.
Electronic document management systems : a portable consultant / Thomas M. Koulopoulos, Carl Frappaolo.
p. cm.
Includes index.
ISBN 0-07-035914-8
1. Text processing (Computer science) 2. Document Imaging systems. I. Frappaolo, Carl. II. Title.
QA76.9.T48K68 1995
651.8—dc20 95-9051
CIP

1 2 3 4 5 6 7 8 9 0 DOC/DOC 9 0 0 9 8 7 6 5

ISBN 0-07-035914-8

The sponsoring editor for this book was Marjorie Spencer, the editing supervisor was Joseph Bertuna, the production supervisor was Donald F. Schmidt. It was set in Palatino by Dina E. John of McGraw-Hill's Professional Book Group composition unit.

Printed and bound by R. R. Donnelley & Sons Company.

This book is printed on recycled, acid-free paper containing a minimum of 50% recycled, de-inked fiber.

To my parents.
Without the technologies of the information age, the world you grew up in was much smaller and simpler than mine, yet your dreams were just as expansive.

—TMK

To my daughter, Anna.
I can only imagine what your world will be like, with unprecedented abilities to access and utilize humanity's acquired knowledge. Hopefully, your generation will use this knowledge to build a world with greater patience, understanding, and love.

—CF

Contents

Preface

Seven years ago, a small group of committed individuals led by two fledgling entrepreneurs placed a bet that document management would become the rallying cry for information systems and business process innovation. They were convinced, as are most entrepreneurs, of their vision. What they could not predict was the frenetic pace of innovation in networking, client/server, imaging, groupware, workflow, and business process reengineering that was to follow.

Today that vision is a reality. The seven years have unhinged the infrastructure of virtually all organizations. Today the reengineering, reinvention, and at the very least, rethinking of the enterprise and its information systems is the most daunting issue facing business and government.

How does document management fit into this upheaval? The common currency of every process is the document. Whether paper or electronic, picture or text, static or multimedia, the document is the essence of communication and the lingua franca of the extended enterprise. As the global workforce turns from laborer to knowledge worker, so does our work turn from producing goods to creating knowledge. The modern day commodity of the skilled worker is authored information—the document.

In this knowledge-based workforce, it is not sufficient to simply automate existing paper-based document management systems. The rate at which we collect unstructured document-based information challenges our traditional methods and technologies far beyond their intended purpose. We must design solutions that redefine collecting, retrieving, and distributing document-based information. Meeting this challenge requires new models for technology and business and a new paradigm for information management.

The *electronic document management system*, or EDMS, is a cornerstone of this new paradigm, which we call the *single point of access*. It is quickly becoming a standard component of every information management strategy and a necessary tool for every knowledge worker.

Electronic Document Management Systems: A Portable Consultant began with seminars and presentations that have delivered that message to more than 50,000 individuals worldwide. This text has a simple objective: to provide the reader with a comprehensive yet simple overview of electronic document management. It is the synthesis of the ideas, concepts, technologies, methods, and diverse range of related issues that collectively make up electronic document management. At times you will find the diversity of topics covered frustrating, since the threads of many ideas weave in and out of this text. There is no easy way to present the myriad topics, and no easy way to digest them. Changing paradigms requires education, time, experience, and often a little pain. The good news is that we are all faced with navigating the same learning curve. By getting this far you have already crossed the starting line.

Both technologists and business people must develop an understanding and an appreciation of the dramatic impact that EDMS will have on their enterprise systems and users. Without this broad foundation, arbitrary expectations will be set, benefits will pale, user acceptance will wane, and yet one more layer of technology with dubious benefit will be added to an already crumbling information systems infrastructure.

Acknowledgments

Since the text embodies the collective knowledge of many years of experience, there were many individuals whose insight and support helped make its writing possible.

Foremost among the individuals helping to make this book a reality was Mary Ann Kozlowski, someone whose perseverance, positive attitude, and relentless capacity to remind us of deadlines was present throughout every phase, from idea to production. She worked single-handedly with the press to position Delphi as the leader in EDMS (long before we were) and was always ready to act as publicist, liaison, writer, copy editor, and conscience. All of our good intentions would not amount to much without her on our team.

In addition we extend our thanks to *each* member of the Delphi team for their help in providing the context of a successful, innovative, and well-respected organization. We have modeled our own enterprise using many of the ideas and technologies proposed in this text. And our own knowledge workers have struggled with adapting to the new workplace that we have so zealously attempted to create at Delphi. That is no small feat. These are exceptional individuals whose character and spirit are the foundation of every successful enterprise.

Thomas M. Koulopoulos
Carl Frappaolo

1

Transformation to a Single Point of Access

Any new technology creates a new human environment. —MARSHALL MCLUHAN

Transformation

Today the information industry is experiencing a transformation as our view of information management changes radically from an isolated task to one that cuts across all dimensions of an organization. With the ability to capture, store, retrieve, and distribute vast amounts of information from individual to individual and workgroup to workgroup, we are pushing the boundaries of existing computing infrastructures and challenging current models of information processing. The result is a paradigm shift away from information systems as an isolated discrete discipline to information systems as a fundamental component of every business and social institution.

This transformation is accompanied by new metaphors and new problems. The most powerful of these metaphors is that of the electronic document. Brought into the limelight by technologies such as imaging, the electronic document is fast becoming a rich new data type that can take virtually any form: text, image, video, or virtual reality. From the user's perspective the electronic document is more than just a new approach to managing information. It is the embodiment of personalized information systems, since it represents the user's own proprietary knowledge base. Documents, from electronic publishing to multimedia, have elevated the intimacy between man and machine.

In this new paradigm, information technology is not only a means to achieve organizational strength but a means to empower individuals. Advances in user interfaces coupled with the enormous decrease in the cost of desktop and portable computing have created end users with a wealth of information and computing power at their fingertips. For the enterprise, this means managing widely distributed and heterogeneous document-based information systems that must not only continue to provide individual value but somehow preserve the value of the entire information base across users, despite its diversity. Accomplishing this will require that organizations reevaluate their current information systems and business processes. A business process redesign that focuses on empowering the user by exploiting the properties of the electronic document is necessary.

This new environment, which we will call a *single point of access,* focuses on a user-centric information system that provides access to all information within one interface—an interface that will make pale the contemporary two-dimensional displays and window-based metaphors we use today. (See Fig. 1-1.)

The full impact that a single point of access will have on today's organizations is virtually impossible to predict. Benefits and drawbacks will arise that cannot be appreciated in the context of current technologies and applications. We can nonetheless speculate as to the changes that will result from this new vantage point and begin to lay the foundation for the technologies, and more importantly the methodologies, we will need in order to prepare ourselves and our organizations for the transformation to come.

We often hear this change referred to in the context of a paradigm shift[1]—but the term has become so colloquial that it has lost all meaning. Each new technology or method is described as a paradigm shift. One wonders, after awhile, how many paradigm shifts can actually occur in a lifetime. Lately it seems that they are occurring on a daily basis. Drastic alterations in a societal perspective do not occur so regularly nor do they occur in the brief span of a few years. If we look at dramatic instances of paradigm shifts, those that have occupied the space of many decades, we can better appreciate the events that today

[1]The term paradigm shift is best described in the works of Thomas Kuhn who popularized the phrase in his timeless text *The Structure of Scientific Revolution* (1962).

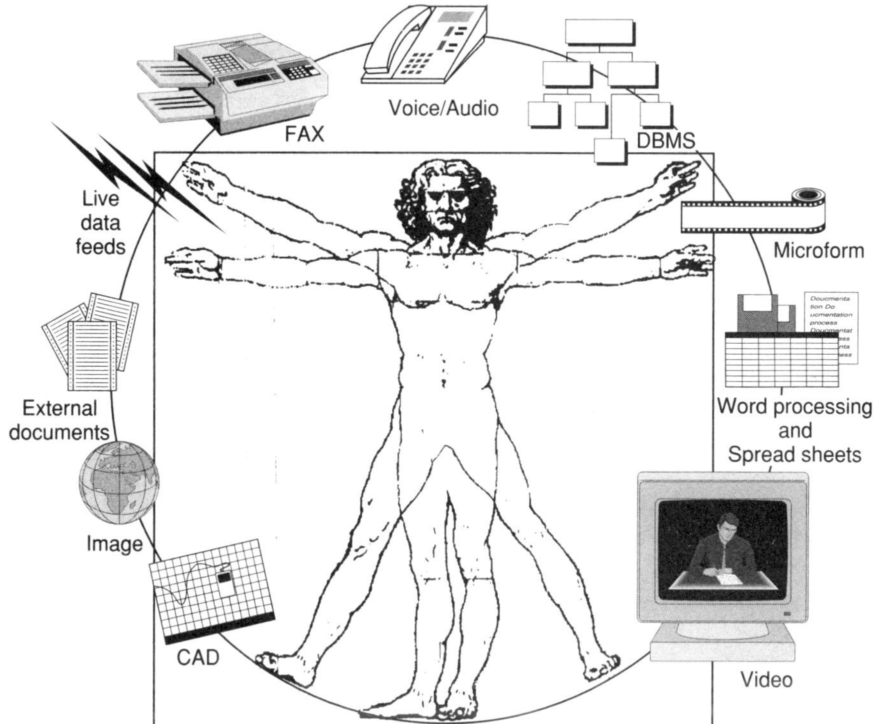

Figure 1-1. Da Vinci man. User-centric information system provides access to all information within one interface.

propel the move to electronic document management systems. The key to appreciating a paradigm shift is understanding the breadth of its implications, which are rarely limited to technology alone.

The best way to describe such an event is to draw a parallel with one that most of us over the age of 30 are intimately familiar. It is, perhaps, one of the most amazing alterations of perspective that we, as a global society, have undertaken—the shift toward ecology. As with any paradigm shift, ecology had a genesis. It is best described in the words of someone who experienced the new perspective firsthand. Recounting his memories of Apollo 14, crew member Edgar Mitchell said, "It's impossible to come around from the dark side of the moon and see Earth floating out there in this vast sea of blackness and not have it affect you in a profound way." Indeed, from the moment each of us first saw that same image of the earth we were empowered by a collective vision and the sudden realization of our fragile and isolated existence. That image, shown in Fig. 1-2, forever changed the way we regard our planet. It is hardly a coincidence that, ever since the space program began and images of the blue planet became com-

Figure 1-2. Image of the earth from space.

monplace, we have undertaken a profound shift toward environmental concerns and adopted the role of planetary stewards.

Such is the legacy of a paradigm. The object doesn't change, the observer doesn't change, but what changes subtly is the perspective. And a subtle change in perspective can evoke dramatic results. The changes occurring today in information systems and the events that are unfolding promise to be just as dramatic.

To understand these changes we should first look back to the origins of the three paradigms of information management: the Black Box, the Blue Box, and the Missing Box.[2] If we trace the genesis of today's informa-

[2]These three paradigms are an oversimplification to be sure. They do, however, represent distinct periods of computing's evolution from a tightly held science to its democratization. Along with that evolution both computers and the information they act on have become open and easily available resources for virtually everyone in the developed countries of the world.

tion systems through each of these, we will see that specific events and phenomena have lead us through three distinct paradigms of information technology and applications.

Three Paradigms of Computing: The Black Box, the Blue Box, the Missing Box

The Black Box

The very first paradigm of information systems is best described as the von Neumann model of processing. (See Fig. 1-3.) Named after the father of modern-day computing concepts, John von Neumann, [3] this model was very linear. It consisted of an input, a process, and an output. It was isolated from other systems. Traversing its boundaries to work with other information systems was not an option or a requirement. The purpose of the process was, after all, singular, such as to calculate the trajectory of a ballistic missile or to tabulate data and extrapolate trends. It was a glorified calculator. The concept of the end user didn't exist in this paradigm, at least not in direct proximity to the process. It became known as the *Black Box.*

[3]John von Neumann was a Hungarian mathematician who first promoted the stored program concept in the 1940s

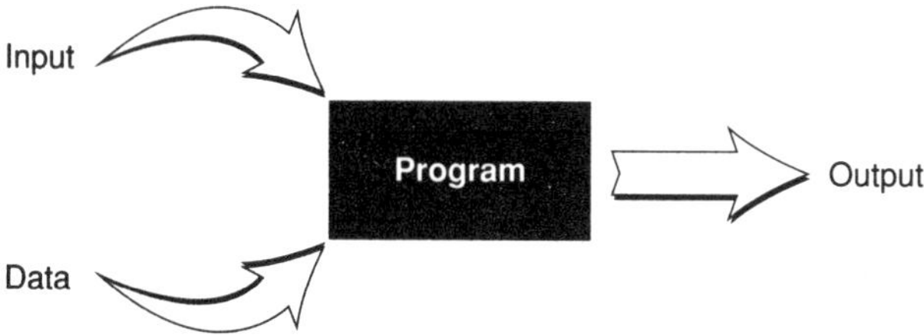

Figure 1-3. The data processing paradigm—also known as the Black Box. The earliest paradigm of information processing focused on the process or the program. Information was dealt with at the lowest levels within highly structured algorithms and registers. User interfaces and report formats were hardly an issue at this stage of the technology's development. This period lasted roughly from the late 1950s until the late 1960s.

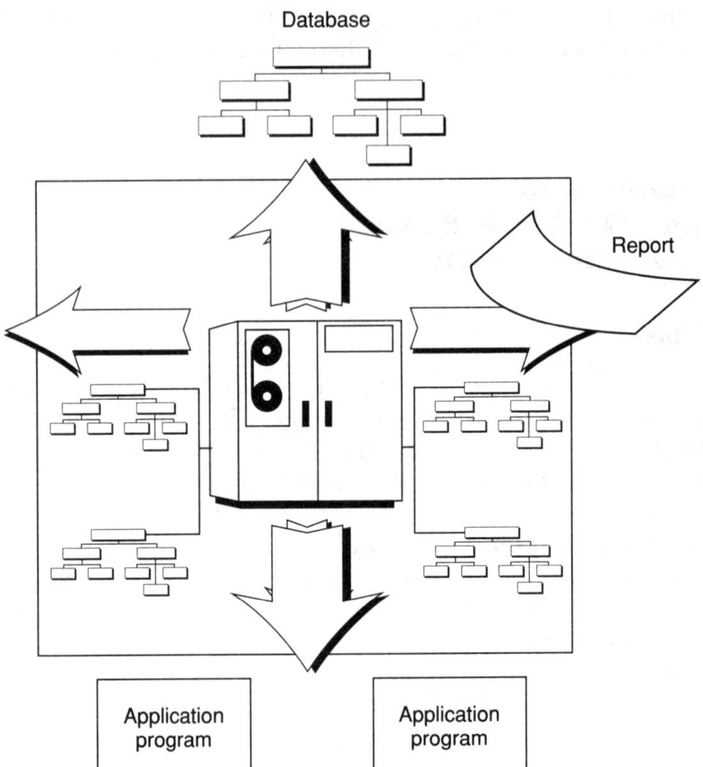

Figure 1-4 The enterprise paradigm—also known as the Blue Box. The second paradigm emphasized the importance of the database manager as the owner and repository of information. Eventually this was expanded to include sophisticated language and programming tools. The application language and the database structures, however, were closely related (i.e., SQL). Furthermore the application environment was still relatively self-contained, with little interoperability.

The Blue Box

The second paradigm, and the one that most of us are accustomed to, was that of the database. (See Fig. 1-4.) The database shifted the emphasis from the process to the data management. From this evolved hierarchical network databases, relational databases, the idea of the enterprise computing with the corporate information repository—epitomized by the large IBM mainframe. Ultimately, from this evolved client/server computing models as the walls around data structures came down. Some filtering was still necessary to traverse those boundaries. Standards were not universal nor were the network connections from one host to another. But the line between the inside processes and the outside world was quickly becoming blurred. Programs became secondary, as did coding. The Black Box gave way to *fourth generation languages* (4GL), CASE tools, and object-oriented programming techniques. This became known as the *Blue Box.*

The Missing Box

In today's information management systems the concept of the central data repository has all but vanished. From the end user's viewpoint, it makes little if any difference where the information is coming from or where it is going. End users simply want to be able to retrieve and transport information. It may be in the form of a facsimile, it might be a scanned document, it might be video, it might be an existing word processing document, it might be a color image, it might be a database, it might be voice. In this paradigm end users require a single point of access through which information of any source and any form can be easily accessed, manipulated, and transmitted. This paradigm, shown in Fig. 1-5, is known as the *Missing Box.*

Although, within this paradigm, the scope of information access has increased and the structure of the information has decreased, the underlying database model has not gone away. That's an important point to keep in mind since many evaluators of EDMS solutions expect them to replace traditional DBMS systems. Nothing could be further from the truth. The fact of the matter is that the database is very much a part of the underpinning of any EDMS. An EDMS that attempts to replace DBMS will typically do so with a proprietary index of its own. That can lead to migration problems as the EDMS grows and interoperability problems as it spans the multiple platforms and networks of a large enterprise.

At the same time we can no longer think of information as simply field-oriented database elements. These new document-based information types are much too robust for the rigid constructs of database concepts. We have to look at this new quantum of information, *the document,* in a very different way, and begin to educate our organizations on the value and nature of electronic documents.

Laying a Foundation for the Future

The challenge for both visionaries and pragmatists is education of the enterprise in the role of electronic documents as a valuable business commodity that translates to a healthier organization and a more profitable bottom line. Education is the key to transforming enterprise

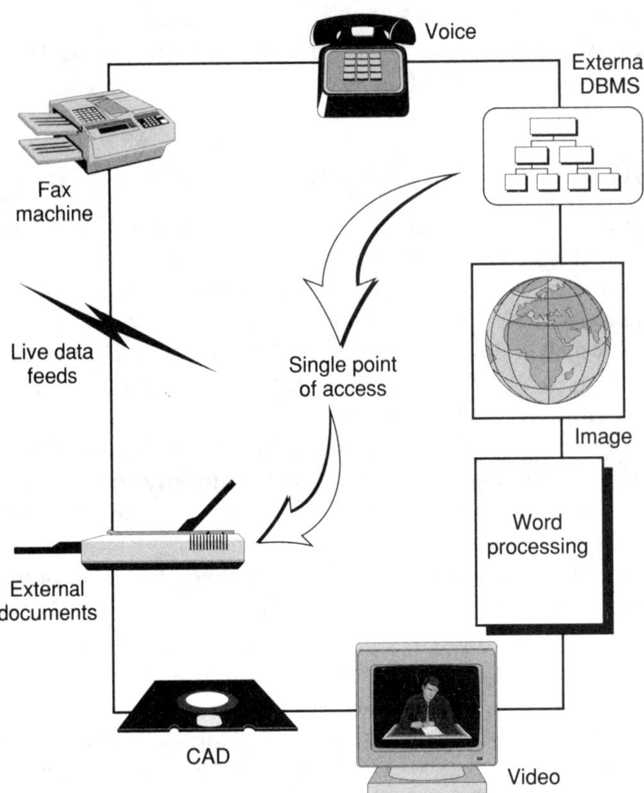

Figure 1-5 The single-point-of-access paradigm—also known as the Missing Box. The third and current paradigm has a greatly expanded focus which regards the database as the hub of an information gathering process, rather than a simple repository. The database is now tasked with managing the collection, interpretation, and presentation of disparate information sources without regard to their origin or format.

information systems to single-point-of-access systems; it seems trivial but the power of education in helping people cope with a paradigm shift, cannot be underestimated. The organization must strive to foster technical literacy throughout its work force and to create an IS environment that is user-driven and focused on incremental reengineering processes. Without this most users will be crushed under the ever-increasing burden of information and technology.

Technical literacy applies to all workers including users and IS professionals. This is especially important in applications and systems involving electronic documents. Because the end user has an inherent ownership interest in the documents, IS no longer has the luxury of controlling the information or the applications that manage it. If users are not able to understand the benefits, drawbacks, standards, and alternatives of the technology, they will fall back to old ways of managing their information, or worse yet, choose solutions that cannot be supported or integrated with other enterprise systems. User buy-in requires this baseline education. Circumventing it undermines success from the outset.

You may want to go a step further and set up some basic standards by which to govern the interoperability of the underlying information in each individual's or workgroup's repository. The objective is to provide a stable footing for the selection of a particular solution and to avoid the use of proprietary solutions. Be careful, however, not to eliminate the user's ability to choose and participate in the development of the solution. Users should have as vested an interest in the solution as they do in the information it is managing.

With that foundation in place you should establish an education in the actual use of the technology. This means choosing tactical applications that can serve as role models for the organization and leverage points for justifying future implementations. This approach may seem to fly in the face of the much ballyhooed obliteration of existing systems or the top-down approach to reengineering, and to a degree it does. Despite all of the hyperbole about finding solutions that meet business needs instead of technology objectives, the two are not exclusive. Implementing the technology is a necessary part of reengineering business processes.

Long-term solutions require vision and a life-cycle approach to reengineering information management systems. The common misconception that we should purge today's inadequate information systems in favor of an ideal solution will only perpetuate the cyclical nature of technology obsolescence and the enormous reinvestment that has come to typify computing over the last three decades.

Unfortunately reengineering has been misinterpreted as a radical process of decimating and then rebuilding information systems. But this ignores the enormous economic and cultural obstacles that many organizations face in adopting even a moderate degree of change. A long-term vision of the organization, exclusive of underlying technologies, is worth establishing but the implementation of that vision will have to take place at a rate of absorption that is commensurate with the existing attitudes and infrastructures of the organization. Over a period of time, these infrastructures, attitudes, and indeed the business itself can be changed. But the available technology will change as well. That creates an inevitable quandary for the "all or nothing" approach to reengineering: by the time we reengineer, the technologies we used or planned to use are outdated. Now reengineering becomes a project to be accomplished every 5 to 10 years rather than an ongoing activity of the organization.

This is where a technology like workflow can provide a balance between short-term results and long-term vision. Workflow does not require radical change in the organization. In fact, most workflow applications are facilitated by the use of an incremental or pilot implementation of the technology. This provides a tremendous amount of experience to the implementing organization. It also shows immediate results, which, in turn, provide a leverage point in justifying future enterprise applications.

Workflow is also one of the most significant advances in electronic document management and the movement toward single point of access. Workflow applications span multiple workgroups, platforms, and technologies and tie together all of the information-based components of a business process. This addresses the problem created in office automation environments where documents flourish rampantly through networks. Workflow allows office workers to exchange, route, and create rules for the processing of information, including images, documents, structured data such as a spreadsheet, or any other form of electronic information. Workflow acts as the information process manager insuring that the integrity of a single process is maintained regardless of the underlying technologies and hardware.

An incremental approach of this type based on established technology standards will provide an invaluable education, show immediate and leverageable results, and empower users with a collective vision of the new system. One of the hardest things for many organizations to accept is that it is the users who will ultimately define the benefit and new work models—not management or IS. If you provide the necessary education and a track record of incremental success, users will become your greatest asset to a successful implementation of single point of access.

After the Transformation

What might the resulting landscape of the information management industry resemble in 10 years when we have made it through this radical paradigm shift? We can only speculate as to the potential consequences and effects.

Matthew Koll, President of Personal Library Software, developers of advanced document retrieval technology, recalls a question he would ask of students in his computer science class at Syracuse University:

"If you had a computer of infinite power and infinite storage, where responsiveness was always instantaneous, what information management problems would remain to be solved?"

It is something worth pondering. Until now our efforts have dealt primarily with the hardware limitations of information management—power has been the measure of value and quantity has been the measure of effective access. What happens when the user is brought face to face with unlimited amounts of information and computing power?

In 1945 Vannevar Bush, Science Advisor to President Franklin D. Roosevelt during the Second World War, wrote, "Of what value is all of mankind's accumulated wisdom if you cannot extract knowledge from it?"[4] He was speaking of the millions of documents and discoveries of the war stored in warehouses. Tomorrow we will have technology that allows us to store all of that, plus the accumulated research of all the generations that have followed, on the tip of a pin. The paper is gone but the problem is much worse.

We talk about the information age with unbounded arrogance, as though it has come and gone. Yet 95 percent of our information still ends up in paper format. When we have learned how to not only store but actually use that information in electronic form, we may finally find ourselves at the threshold of the much touted information age. That's when the real problem of information management will begin.

With the power to liberate knowledge from virtually any information base, each of us will have the opportunity to become an expert of choice in areas that would otherwise be beyond our limits. The era of specialists will give way to the era of the generalists, where individuals and organizations are no longer at the mercy of "experts." Information is available to all who have the tools and the knowledge to access it. What will happen to the specialists? the same thing that happened to the craftsman of the last century. Some will remain, but the revolution of information accessibility will cause the automation of the information gathering and retrieval process in the same way that the industrial revolution caused the automation of the manufacturing process. This may finally deliver the productivity we have promised for so long. Along with it will come the displacement of white-collar workers and the inevitable segregation of the information rich and the information poor—that may be the greatest societal and competitive rift of all.

But let's not stop there. What might the organization itself look like? Sophisticated workflow may replace middle management. Researchers and analysts will be replaced by knowledge agents and information

[4]V. Bush, "As We Think," *Atlantic Monthly*, 1945.

refineries. Administrators will be replaced by intelligent documents. Librarians will be replaced by judgment-based retrieval software. As for the paperless office, the office itself will be replaced with single-point-of-access personal information appliances that can tie into any information located anywhere. Single point of access will ultimately become as individualized as the wristwatch. Perhaps the skyline of tomorrow's cities will be littered with the remains of empty office buildings, their former inhabitants no longer needing to go to work because the work goes with them.

This may seem far-fetched, disquieting, even anathema to some. But such is the nature of paradigm shifts. Each age of humanity has been just as difficult to fathom for its preceding generations. Every so often the world changes in dramatic ways, which are usually as obscure in their genesis and as profound in their impact as the view of the earth from the lunar horizon. EDMS has the potential to change information management in such a way.

If you have come this far then you are one of the harbingers of this new era in information systems. You and those who take the challenge and join you will need to understand the principles and tenets of this powerful new set of tools and methods in order to manage the inevitable transformation to come. This text will set forth these principles. It will begin your education and it will set the stage for revolutionary changes in information systems. But remember that documents belong to everyone. They are not the sole domain of the scholar, the scientist, or the researcher alone. In the end it is you who will be the agent of change and its pilot. To that end remember that technology held only in the hands of the power user empowers no one until it empowers all users—each and every one.

2
Document Management: What's the Problem?

A Popular Myth

A popular myth abounds in today's information-rich organizations: we are drowning in paper. Promulgated by an information industry that is desperately trying to reposition itself after a decade of hardware commoditization and falling margins, the cry to eliminate paper is the calling card for a cadre of new technologies, peripherals, and computer platforms. It has also become the focus of attention and planned investment for many organizations trying to cope with overburdened information systems and business processes. If you buy into the myth, then you are likely to accept as fact that simply replacing paper with electronic images is a sufficient solution to the problem. But what if the problem is more than paper? What if there is a deeper set of issues that we should be addressing? It may be that by ignoring the fundamental problems we are simply adding one more layer of technology to an already crumbling IS infrastructure. Before investing in yet another technology solution we should at least invest some time in understanding the problems.

The myth referred to took shape when the problem was first popularized in 1987 by Coopers & Lybrand in a study which has since become nothing short of a cult icon.[1] (See Fig. 2-1.) The study showed that 95 percent of the information available to an organization is stored in paper form. It seemed clear that if there was a problem it must be due, in large

[1]Coopers & Lybrand study showing that 95 percent of information is on paper.

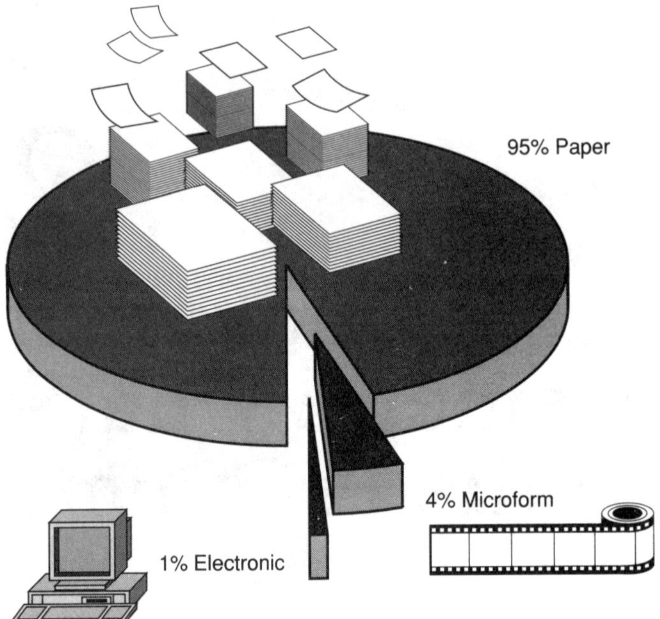

Figure 2-1. Document Management: What's the problem? Ninety-five percent of the information available to an organization is stored in paper form. (*Source: Coopers & Lybrand, 1987.*)

part, to the overwhelming presence of paper. After all, paper requires significant manual effort to manage. It is subject to any number of physical problems and limitations, including misfiling, aging, and single-user access among many others. That same study went on to show that only 1 percent of all corporate information was resident in electronic form. The evidence was difficult to ignore. Computer-based information had very little impact on how organizations were run because, fundamentally, organizations were still run on paper.

Since 1987 that particular study has been immortalized in numerous graphics, product literature, and print media. It's been seen over and over again to the point where, like a popular song, it loses all novelty. In fact there is nothing amazing about these findings. What is surprising, however, is that we have so readily bought into the premise that this is a sufficient problem statement. It is not. It is, instead, the most easily identified symptom.

Defining the problem becomes much easier if we move forward a few years from the 1987 Coopers & Lybrand study. Since that time many technology solutions, such as imaging, have come to market claiming to solve the paper problem. Despite enormous expectations and predictions of stellar market growth, imaging has not proved to be a panacea to the paper avalanche. What imaging does prove, however, is that we may indeed have ignored the more important but much less obvious issues facing information and document management.

A study similar to that conducted by Coopers & Lybrand was conducted in 1991 by Delphi.[2] Its results show clearly that we have been able to chip away at the paper problem. (See Fig. 2-2.) According to Delphi's percentages things have gotten much better. Today only about 90 percent of the average organization's information exists on paper. A relatively large piece of that information base, 3 percent, is in electronic format,

[2]Delphi 1992 study on paper use in organizations.

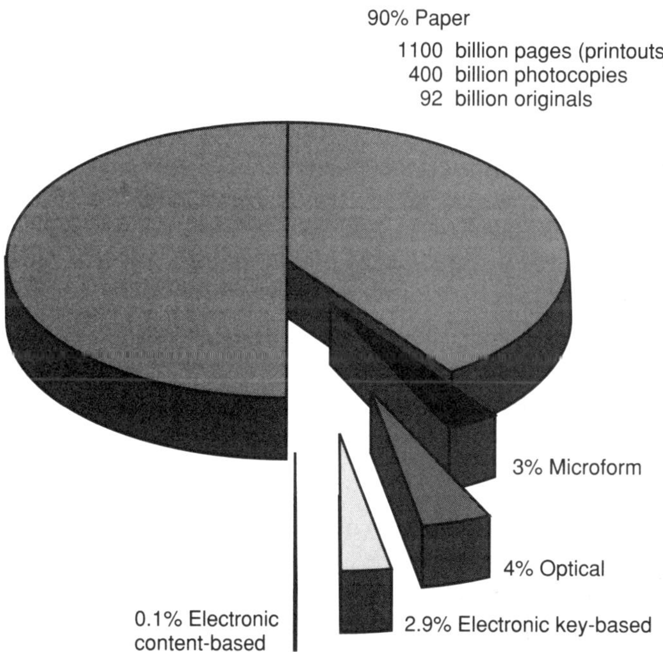

Figure 2-2. Document Management: What's the solution? Dealing with the migration of paper-based information to electronic format requires a broad variety of technologies to help with the storage and retrieval of this information. These technologies, however, are still in their infancy, with only a very small percentage of this information currently stored on electronic or optical media. What we today perceive as solutions to the problem of information overload in turn create a new set of problems and integration issues that must be addressed before industry and society at large can benefit from this wealth of information.

and 4 percent of it is in optical format. This means that fully 7 percent can be accessed by a computer-based application. The remaining 3 percent, a decreasing portion on a relative scale, is stored in some sort of microform.

Has the problem gone away as a result of this progress? Has it at least improved? That is hardly the case. In fact, the problem appears to be getting worse. Managing information does not seem to be a direct function of converting paper to electronic form. Perhaps that's because of the 1 billion pages of paper that we generate in the United States each day, 70 to 80 percent is generated by computers. And of that, 60 percent is used for data entry into another computer-based system!

Ultimately it is difficult to ignore the underlying problem, of which paper is merely a side effect. A pathologic condition exists in information systems today—they cannot share the richness of document-based information. Paper is still the most compatible means of transferring documents and their contents from one island of technology to another.

Islands and Bridges

Throughout the 1980s we often spoke of *islands of information,* a phrase popularized by McKenney and McFarlan in their 1982 *Harvard Business Review* article "The Information Archipelago—Maps and Bridges."[3] The fact is that the monolithic corporate and departmental systems of the 1970s and early 1980s were better described as *continents of information.* In contrast today's IS infrastructure is a heterogeneous, highly distributed, user-centric environment. Organizations are rich in information if only they could bring together the diversity of platforms, data repositories, and applications. It is little wonder that the common denominator for so many computer-based applications continues to be paper. With such disparity among information systems, paper has remained the only constant and dependable means of transferring and sharing information across technology borders.

The obvious cost of paper creation, retention, and retrieval aside, the impact this has on reduced productivity, quality, and responsiveness is a liability to every organization. Perhaps the only beneficiaries are the manufacturers and suppliers of laser printers. Their success has been a powerful testimonial to the ineptitude of different computer environments to communicate electronic information read-

[3]McKenney and McFarlan, "The Information Archipelago—Maps and Bridges," *Harvard Business Review,* 1982.

ily from one electronic repository to another. The information archipelago has indeed arrived and the bridges from island to island are paved with paper!

The problem with implementing an electronic document management system to handle this, however, is that document-based information is a proprietary and highly personalized resource at the individual, workgroup, departmental, and enterprise level. As a result electronic document management systems are springing up in every nook and cranny of the organization as fragmented and often dedicated applications—furthering the problem of isolated information systems. As the cost of desktop computers and peripheral devices such as scanners and optical storage continues to fall, the problem will only get worse. In fact, today 95 percent of all electronic document management systems are available on PC platforms.[4] Many of these offer powerful solutions for only a few hundred dollars per user.

In this avalanche of paper it is easy to see why imaging has been such a popular technology, but the obvious benefit of paper reduction is only part of imaging's contribution to transforming information systems. The greater role imaging has played is in opening the door to a common electronic metaphor by which to join otherwise separate and distinct elements of the organization—the electronic document. That shift to user-centric, document-based information is the foundation of technologies such as workflow and groupware. By evaluating or implementing imaging, many organizations have been forced to look carefully at these technologies as well as the overall problem of accessing and sharing not only images but all types of document-based information across workgroups.

What organizations implementing imaging are realizing is that imaging alone has some benefit but it does little to change the basic work process and task metaphors. That, too, should come as no surprise. If we have learned anything from the last two decades of rampant technology innovation, it should be that technologies must coexist in an integrated environment. Despite promises of a paperless office, imaging has not replaced paper or microfilm. Both are still here and both will be viable information management alternatives for some time to come. (See Fig. 2-3.)

The question we should be asking is not whether imaging, or any other single technology, is the ideal metaphor for information management. Rather, we should determine how imaging and other document management technologies such as workflow, text retrieval, E-

[4]Delphi Imaging Report, 1993.

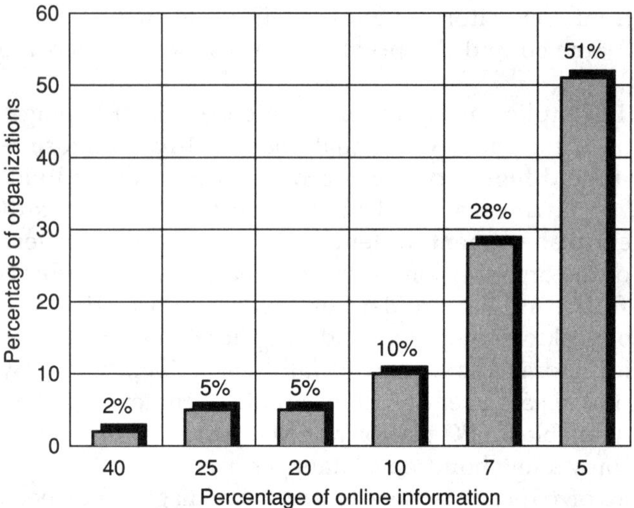

Figure 2-3. Document management: What's the solution? The trend is clearly toward electronic document creation, storage, and delivery. But how far along are we on the learning curve? According to Delphi's studies some organizations have made significant progress. In fact 5 percent of all organizations have 51 percent, or more, of their corporate information on-line in electronic format. The vast majority of organizations, however, still have no more than 10 percent of their information on-line.

mail, word processing, database, and a myriad of others, including existing legacy systems in paper and microform, come together to create a single point of access for all users throughout the organization. That cannot be determined through the application of technology alone.

Ultimately paper and many of the technologies we have used to eliminate it point to a much deeper problem, one that results from our inability to transfer electronic information across the vast variety of hardware and software platforms in the organization. The last two decades have been a frenetic attempt to invest in a succession of latest and greatest technologies. The lesson we have learned, if any, is that the information management problems of an organization cannot be answered through the application of any one technology. Even in a utopian setting, with all of the technology and economic barriers removed, we will still have to face up to the greatest challenge of all—transforming the organization and its people instead of just taking another turn through the constantly spinning turnstile of technology obsolescence.

Information Gridlock and the Glass Desktop

The phenomenon that everyone seems to be experiencing today is that of information gridlock. Information seems to congregate in its lowest common denominator, the paper document, where it is ultimately drawn to the only universal form of information integration available in most organizations—the desktop. The enormous misnomer of the computer as a desktop becomes a bold-faced lie when you consider the inability of any computer monitor or present-day interface metaphor to mimic the variety and richness of the information that is presented on a 3- by 6-foot physical desktop. It's a wonderful metaphor, I agree, to think of the screen of the PC as a desktop. But the problem of gridlock is based in large part on the inability of present-day metaphors to replicate traditional work environments or to provide sufficiently novel and beneficial alternatives. (See Fig. 2-4.) An assignment at the Mayo Clinic revealed the reality of this problem.

The Mayo[5] Clinic had undertaken a study of its records management procedures in an effort to convert from a paper-based system to an EDMS. Historically, patient records have been kept in plastic envelopes. A single envelope contains at least a few dozen pieces of paper and may have more than 100. The patient records are shuttled back and forth from doctor to doctor and to records retention areas through an elaborate series of complex conveyor belts, chutes, and pneumatic tubes, some of which span more than a mile of the facility.

At the Mayo Clinic doctors examine patients in their offices. The dimensions of each office are roughly 10 by 20 feet, with a single modest desk and bookcase in one corner. Every doctor at Mayo has the same size office. All Mayo furnishings have been custom-built to the same rigid specifications for consistency in ergonomics. The paper is annotated and reviewed by each examining physician during the patient's examination. According to one analyst of the system, these records are used in such a way that doctors form a neural relationship among scattered pieces of paper littered across their desktops. This practical element of existing systems is an undeniable aspect of diagnosing a patient's illness. One of the principal problems faced by Mayo in moving to an EDMS was the lack of desk space for large-screen monitors that could adequately represent the wealth of information contained in a patient's folder.

[5]The 103-year-old Mayo Clinic, located in Rochester Minnesota, may have arguably one of the most elaborate and efficient applications of manual document processing in existence. Originally designed by a visionary, the system combines a variety of manual and computer-based systems to process 1.5 million patient records.

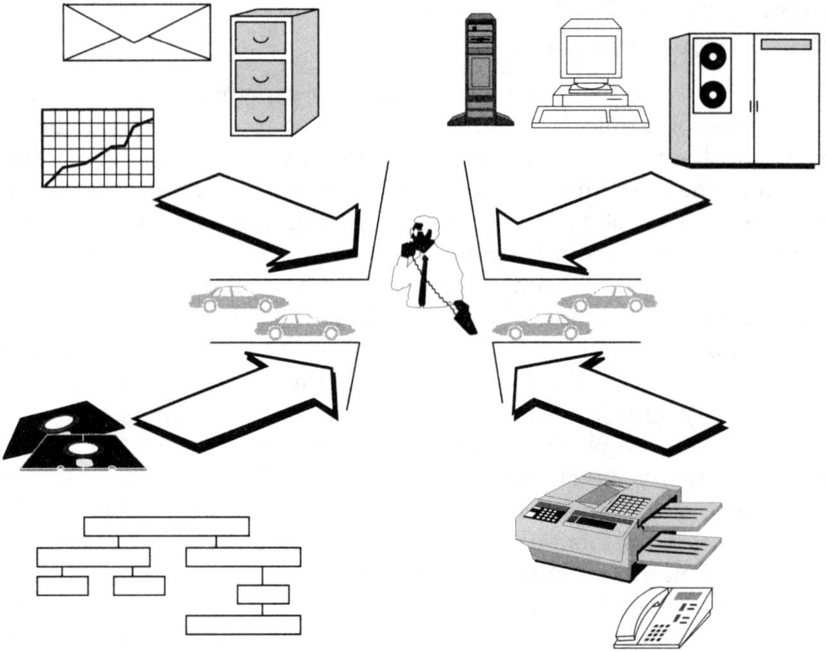

Figure 2-4. Information gridlock. Today's organizations must deal with a myriad of information sources and types. In many cases the result is frustrated knowledge workers who cannot retrieve the information they need and are drowning in information they don't need. The overwhelming amount of information available to today's information worker is dramatically reflected in the hours spent, on average, searching for misplaced information. Estimates place this at 150–250 hours per year for the average executive and 500–750 hours per year for a manager. The time spent searching is both unproductive and costly. Minimizing this effort can result in tremendous productivity gains and cost reductions relative to paper-based information systems.

One alternative which may soon be regarded as technically feasible, was the creation of a glass desktop. In this model the entire desktop would be replaced by a high-resolution monitor. The doctor could move electronic documents around the glass desktop, rearranging each electronic document in the same way the paper document would have been handled. Electronic baskets for incoming documents, filing systems, mail, and other facilities would be represented on the glass desktop. This metaphor would not require additional space and would parallel the existing approach to managing documents.

Unfortunately today's technology is not quite able to provide the glass desktop. Until it is we will undoubtedly have paper to deal with. But this is not, in and of itself, a problem. In fact, we should redefine our concern about paper production and usage in information processing and document management. Paper is an ideal and often-preferred presentation format. Its portability, low cost, and ease of use are in many cases unmatched by technology alternatives. Rather than attempting to eliminate paper we should simply attempt to use it where it is best suited, as a presentation format of information but not as a repository of information. If recycling is encouraged and paper is used in this basic capacity, it fulfills a necessary and cost-effective role in document management, a role which may not be superseded by any technology alternative for the next century, and perhaps even beyond.

Solving the problem of gridlock lies in identifying where it still makes sense to use a paper-based management system or a traditional DBMS, and where it pays to use an EDMS. This is best understood by defining one of the key liabilities of structured data management systems—the process of distillation.

Data Distillation

The principal reason that traditional data management systems have been inadequate for the management of documents is the structure that they impose on information. A basic tenet of any EDMS is that structure always equals distillation. *Distillation* is the process of eliminating, summarizing, or in some other way reducing a body of information to its essential components. In the case of a document, however, the essential components may change from use to use or user to user. When a document is distilled by imposing structure upon it, you eliminate part of the document's information value. It may not eliminate its value entirely, but it will at the very least limit the future access paths to the document. (See Fig. 2-5.)

A simple exercise helps to prove this point. Select a document and jot down a few short sentences that represent its essence. Now come back to it a few months later and try to go through the exercise once again. If you compare the two summaries, chances are very good that they will have some significant differences. The document hasn't changed. The words and pictures in the document are identical to what they were when you first read it, but your perception, your environment, and other external factors may have changed, causing the essential elements of the document to change.

Distillation results from:

- Limited indexing

Figure 2-5. The problem with traditional IS. Information distillation is the process of extracting pertinent information from its context in an effort to conform to the limitations of the database structure. In many cases this process undermines the value of the information management system by filtering out information which may have value at a later time. EDMS is intended to prevent this problem by preserving the original document in its entirety. The issue now, of course, is searching a larger base of information and dynamically organizing the results for presentation.

- Lost content
- Structured access

In each case the information after distillation is far less valuable than it was in its original form. The irony is that the original form of information, or what we will call *true form* information, is not often available through an EDMS. Although computers certainly house much of the information used to publish and produce paper-based output, this same information has not been accessible through a document storage and retrieval system. In fact, the robust nature of published documents and the very features of design and layout that make them so meaningful in paper cannot be supported by many electronic storage and retrieval systems. Until such solutions are widely available, distillation will continue to be an obstacle for evaluators and users of EDMS.

Some of the most recent approaches to electronic paper use viewers that allow for the transport and display of published documents across multiple platforms. These approaches still lack the ability to provide content-based document management facilities. In short, the problem is not just one of storing documents but more importantly of managing and retrieving them easily from vast electronic warehouses.

Appreciating the peculiarities of an EDMS and the value which it can bring to an organization first requires an understanding of information distillation. Distillation is a necessary by-product of any repository which imposes field-level structure on database elements. All structured information bases perform the process of data distillation in order to impose structure on the information they contain. The problem with distillation is twofold:

1. *A static set of rules must be imposed on the sources of information in order to separate the valuable information from that which is considered superfluous.* The rules must be static so that they will not create an inconsistent distillation process, and also because structured databases must have an ordered and defined set of relationships among data elements. This approach provides a severely limited means of access to documents.

For example, when a major oil tanker ran aground recently it became imperative to capture and store all of the media press about the incident for future reference in any litigation activities.[6] The information was robust free text consisting of small news clippings, pictures, photographs, and audiovisual reports. The distillation of this information was conducted by a group of individuals dedicated to the process of reading, viewing, or listening to each relevant media source and extracting key words or phrases which were input to a structured DBMS. The original source of information was sequenced by a unique key and filed away for future reference.

When information was needed, the DBMS was used to identify the documents pertinent to the incident by the use of these keywords. Consider for a moment that the individuals entering the distilled information may not have had a set of rules broad enough or timely enough to identify all of the relevant words and phrases required for adequate retrieval. Perhaps the specifics of the incident were not fully appreciated when the distillation rules had originally been developed. As a result, future research against this document database will yield results that reflect the prevailing attitude about the incident at the time the article was abstracted. Instead, it should provide results that

[6]A hypothetical situation.

are relevant to the changing nature of new research and findings. The problem was a document database frozen in time and limited in its utility due to a static distilled representation of each document's content.

Whatever the reason, and there are many, distillation is always an incomplete representation of the original information source.

2. *Distillation not only eliminates access to the original but it also provides a false sense of completion to researchers.* Structured databases create a confined research process. The parameters of the research are defined by the distillation process, and in effect some of the potential answers to any question have already been eliminated by the imposition of extraction rules that identify what is important and what is not. Library systems are a particularly good example of this. Information in a library is indexed by a standard set of criteria: author, subject, title, date, and classification. Access to information is thereby limited to items which are easily distilled from the original.

This method has, in fact, defined the tedious nature of research for hundreds of years. Since the method of access is limited, a researcher who wishes to conduct a thorough analysis must either plow through every reference source, even those only remotely relevant to the subject, or limit the research to those reference sources identified as immediately relevant by the existing access paths. Practicality usually dictates a compromise between the two, but often favors the least difficult and time-consuming method—using existing access paths. In the final analysis we expect the research to be complete when we exhaust the existing access paths. We are, in fact, usually far from completion based on the available information. We have simply reached the limits of structured access.

Dealing with Distillation

Solving the distillation paradox of using structured access to achieve limited results, and using unstructured access to produce too much information, requires not only access to the original but also requires the ability to dynamically filter the original. Distillation is still required for effective research, but the difference in the point-of-research EDMS model is that the distillation occurs at the time of query and not at the moment of input. This filtering process is the fundamental building block of EDMS. The filter must combine the original information source, the query process, and the intelligence of a dynamic rule set to extract relevant information from an unadulterated document database.

Advanced implementations of text retrieval technology are an excellent example of how distillation is eliminated from front-end processes and preserved as a research-oriented task, performed dynamically during the query process.

This aspect of distillation is a critical part of a successful document management application. If the distillation is performed correctly the precision and recall of the application are both optimized. If it is not both will suffer.

Three distinct approaches are taken by existing document management vendors today to minimize distillation and optimize precision and recall. Each approach has its advantages and disadvantages:

1. In the first approach distillation is minimized by storing the original document with redefined access paths through a standard database structure. This approach is typical of imaging systems that use a DBMS or a file folder metaphor for the organization and retrieval of documents. At the time of submission, each user query is evaluated against the database, which typically contains keywords or abstracted versions of the document. This process is called the *keyword method of retrieval* since every document must be investigated by someone, or by a predefined set of word lists, when it is added to the document database in order to extract the keywords for inclusion in the database. Depending on the complexity and the dynamic nature of the documents this can be both time-consuming and excessively restrictive. The benefit is that the documents now become part of the database application to the extent that they can be retrieved through a standard database query, such as SQL.

2. In the second approach distillation is not performed directly on the document but instead document information is extracted and organized in a separate data structure to be used as an index to the original. This process is called the *full-text method of retrieval* because of its reliance on specific words and phrases. Full-text retrieval preserves the original document and access to its contents without predefined access paths; however, the index acts as a form of distillation by limiting access to specific words and word combinations. It does not consider the concepts or ideas that may be embodied in the document.

3. In the third approach documents are not distilled through the extraction of information directly or into an index; instead, they are distilled through the process of clustering. Clustering groups of like documents together logically in such a way that when queries are processed they result in the retrieval of document collections that have similar meanings and, as a result, a relevancy. In practice the clusters

contain a variety of documents with slight variations in their relevance to the query. Unfortunately there is no method to date of organizing these clusters without a high degree of human arbitration. Several heuristic methods do exist for this purpose but they are less than perfect. Since they rely primarily on the co-occurrence of words, character strings, or patterns, anomalies may occur if the document is poorly written or if it uses excessively flowery and verbose language. This may be particularly true of a very heterogeneous document set.

In each of these cases distillation is not eliminated but refined. It is still an important part of the process of retrieval, and until such time as the computer can assume the responsibility for inferring meaning without imposed structure, it will continue to be necessary. The key to developing a correctly structured EDMS is defining the exact role and execution of the distillation process—not in eliminating it.

There are three methods for accomplishing this goal: preserve the structure of existing paper metaphors, develop a knowledge-based retrieval method, and develop research skills and tool sets for on-line retrieval. The objective of all three methods is to manage documents in such a way that their original form is preserved, thereby minimizing distillation. It is important to note that these options are not exclusive. An effective EDMS combines these methods in order to address the retrieval, presentation, and usability of the true form documents.

Imaging and Distillation

The first method, preserving the paper-based metaphor, is exemplified by most imaging systems. The document appears to be left intact with absolutely no distillation. The irony is that in many ways imaging alone is perhaps the most devastating form of distillation since it completely eliminates the ability to retrieve the document in any manner other than key-based retrieval.

This is not to say that preserving the paper-based metaphor is an incorrect means of managing documents. It is a key component of effective EDMS, but only when coupled with a method of retrieving the document based on its full content. This includes all data objects that may be part of the document. The easiest of these objects to work with is the document text; however, pattern matching technologies based on binary representations may also be able to retrieve based on image and audio objects embedded in the document. The key, once again, is the ability to work with the paper-based metaphor as more than just a single entity by accessing its components at their lowest level of understandable detail.

Knowledge-Based Distillation

The second method, developing a knowledge base, focuses on the value of the information as it applies to the context of a particular subject or discipline. The concept behind this approach is that the distillation of documents occurs not only at the level of detail embodied by the characters and the objects in a document but also in the implied meaning of the document. This meaning can only be ascertained by considering the document's rank against certain predefined categories of information. The problem, which is immediately evident, is that predefined categories create distillation. Although this is true of retrieval based on the categories defined in the knowledge base, it is not true for retrieval based on the documents themselves. Knowledge bases do not impose point-of-entry distillation on documents. Instead they provide a layer of access that goes beyond the document. In this way they augment access to the full information contained in the document with access to information implied by the document. The key is to provide both levels of access in order to avoid point-of-entry distillation.

Human Factors and Distillation

The last method, improved research skills, may be the most important of all tools to minimize the downside of eliminating distillation. Despite the benefits of eliminating distillation, there is the constant problem of overwhelming the user with information resulting from poorly executed queries on large document collections.

The results of numerous studies on the query habits of users demonstrate that most users are reluctant to use more than two to three words in their queries, and rarely use sophisticated boolean or iterative queries. In these cases users are apt to feel frustrated by the apparent failure of the EDMS to find exactly what they are looking for.

The inverse can be just as dangerous. Users whose overly complex queries result in only a few documents may have a false sense of security in thinking that they have exhausted all relevant occurrences of documents that satisfy their query. Effective distillation must be managed at both the point of entry and at the point of research. This requires a combination of retrieval tools and training on the use of these tools.

When all three of these methods, the paper-based metaphor, a knowledge base, and research skills, are included in an EDMS solution users will have the opportunity to transition quickly into this new information environment and readily appreciate its benefits.

What Is a Document?

We are going to define a new quantum of information and we will call that new quantum of information a *document*. Easily said, but the word document alone is ripe with ambiguity. It can be used as a noun or a verb.[7] It can be a page or a volume. It may be rich with meaning yet defy categorization. Perhaps it is this very ambiguity that lends value to documents. The document is a collection of information, authored for the purpose of transferring and preserving knowledge. Its manifold forms allow it to be a universal metaphor, information repository, and vehicle for knowledge dissemination.

The first rule of the document that we must agree on is that a document has little to do with paper. A piece of paper is only one way to represent a document. It is a metaphor that we are all very familiar with, but it is also a very limited one. So the very first thing we must do is strike from our minds the idea of a document as a page of paper. Figure 2-6 reinforces this change in definition. When we talk about documents and pages, we are speaking of two separate things. A document could have a hundred pages, a thousand pages, or just one page, but that's only one option for a document.

The whole concept of the page may give way to new and novel methods of organizing information. The problem most of us have, however, is that we can buy into that concept but we have no clear sense of what the new metaphors or methods may be.

In a study we conducted for IBM of several hundred users of electronic documents we asked two questions that exemplify this

[7]Think of the statement, "I have documented [verb] the conversation for future reference," as opposed to the statement, "I have a document [noun] of the conversation."

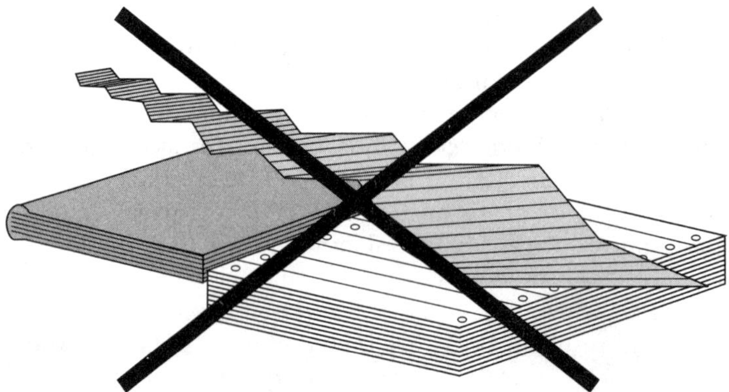

Figure 2-6. Strike this idea.

quandary.[8] One asked respondents to indicate which metaphor was most appropriate for the use of electronic information on-line: the page, the paragraph, the chunk, or some other as yet undefined metaphor. Ninety percent indicated that it was some other undefined metaphor.

The second question, presented later in the survey, asked how many would still want to be able to print on-line information for use in paper form, implying the use of the paper metaphor such as the page. Ninety percent said yes, they would want to continue printing in paper form. It is not a contradiction as much as it is a feeling that although the paper metaphor is a safe and reliable one, there must be a better way to work with the information on-line, if we only knew what it was.

If we strike the idea of the page as the metaphor for document, then how do we define document? There are parameters that we can use to bound the document, but I will caution you that these are ambiguous and it takes a little time to fully grasp the concept of the document in its new context.

The document does contain certain information entities such as image, text, video, or voice. But we should not limit the document to those information entities. If we talk about virtual reality, documents could contain tactile stimulation. A document could contain taste and smell. Each of these and any other sensory input you can imagine is a valid document information type. (See Fig. 2-7.)

Let's start bounding this a little more. Take a typical document collection. (See Fig. 2-8.) Documents can consist of images and text and certainly many other information types. Let's look at images and text to begin with. Assume we have a document collection that contains contracts. That document collection actually contains two basic forms of information: uncoded and coded information, the difference between the two being that uncoded information does not have the ability to present itself in a granular fashion. It is entity-based, meaning that the image exists only as a complete image. There is no way to look at an image of an automobile, for instance, and to retrieve it based on the fact that it has a particular fender or wheel style or, for that matter, any visual attribute unless I have represented that information in a structured environment such as a database.[9] The only way that I can get to the document is as an entire entity. It can't be decomposed any further and as a result, there is no level of granularity available for

[8]IBM study of on-line documentation uses and users.

[9]Techniques to retrieve images based on their content are in development. The basis for these techniques is found in machine vision applications and pattern recognition algorithms. They are not commercially available at the time of this writing but are likely to be incorporated into EDMS solutions by the time this book is published .

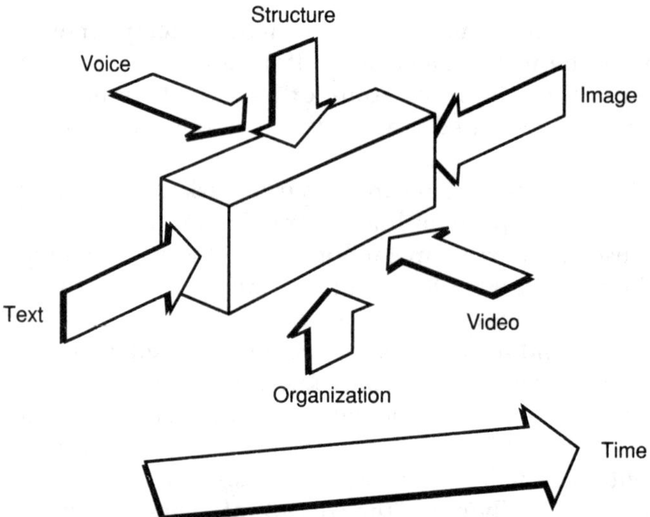

Figure 2-7. What is electronic document management? Electronic document management requires the ability to work with a variety of document formats which may not always be anticipated by developers of an application or its users. Images, text, and graphics can all be stored as images. Based on the usage of the information, however, this may or may not be the most efficient or effective method. The document type, method of retrieval, and storage requirements will determine the content and architecture of the document and the resulting information technologies needed.

retrieval purposes. This can only be achieved if descriptive information is added to a structured index or database linked to the image.

This additional information is authored through manual or semi-automated means, which we will discuss in Chap. 4. For the time being it is important to remember that the authoring of the image index will require significant time and effort, in many cases resulting in the major cost of converting a paper backlog to electronic images.

The retrieval facilities used for a document collection will always take one of three forms, a database, content-based retrieval, or structured retrieval. The database approach will act as an external index to the document, as is the case with imaging. Content-based retrieval will actually use the words, ideas, and concepts expressed in the document. For example a full-text retrieval system will index every word in the document for retrieval. Structured retrieval is a hybrid of the two. It is best characterized by markup languages such as *Structured Generalized Markup Language* (SGML), which will tag sections of the

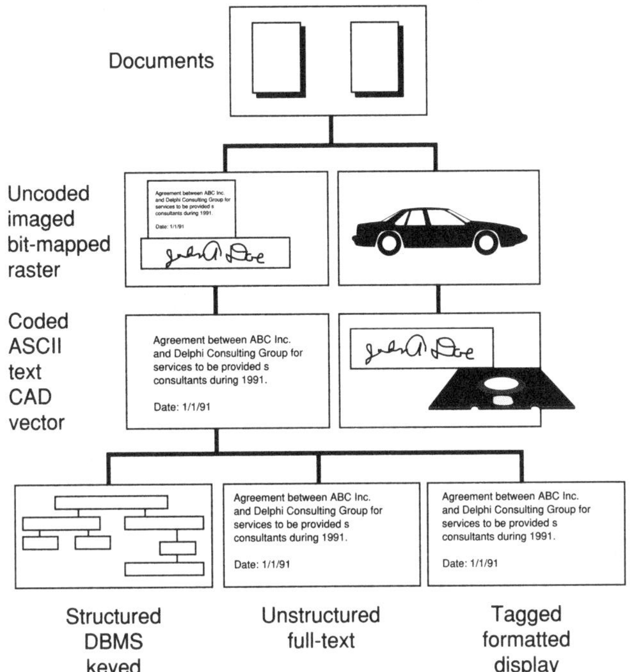

Figure 2-8. What is a document? Most document-based information can be categorized in one of four ways: (1) CODED information which can be interpreted into atomic components by the computer-based system, for example, ASCII or EBCDIC, or CAD vectors. (2) UNCODED bit-mapped information which is simply a series of 1s and 0s arranged to display a graphic or image. Textual information may be stored in this fashion but it cannot be directly interpreted into ASCII or EBCDIC without character recognition software. (3) STRUCTURED information which is field-oriented and contained within a database file. (4) UNSTRUCTURED coded information which cannot be stored in a field-oriented database, usually alphanumeric documents of variable length. Some documents may contain more than one of these types of information in which case they are referred to as compound or mixed-media documents.

document to identify each as a structural component of the document. We will look at each of these in more detail but for now let's consider the issue of structure.

Information type alone is a limited means of defining the document, since any combination of information types—coded, uncoded, structured, and unstructured—can compose a document. A better way to arrive at a definition is to look at the document as consisting of information as well as structure and organization, and there *is* a difference between structure and organization.

The structure of a document relates to two aspects of the document. The first aspect of structure is the way the document is laid out and structured for consumption. In other words, if you look at a printed document, the layout of that document should help you to understand and easily assimilate its content. Why? Perhaps the font styles happen to be particularly attractive, or the layout of the document happens to be particularly suited to you as an end user. The more complex the document, the more important structure becomes to the document's layout and presentation.

The second aspect of structure is the relationship of each item on a document to each other item and to the overall document. In other words, the footnotes on each page of the document are consistent, and they occur in the same location and use the same basic format each time. The titles of chapters, a subhead, call outs, and page numbers are all consistent and predictable. This consistency allows you as a reader to better work with the complexity of a document. It also structures the document for purposes of segregating the various pieces of information it contains. This provides a means to navigate quickly through the document, develop an index or table of contents, or simply identify relevant content.

Organization is the flip side of structure. Organization defines how various documents relate to one another. How do I go from one document to another? What is the link between the two? Is it a logical link? Is it an authored link? Is it an automatic link? Is it based on an index? Is it based on some sort of hypertext functionality? Is it heuristic in nature? Is it knowledge-based? Organization allows us to answer these questions because it provides linkages among documents in a collection. Organization may also be based on the document structure. For example, numerous documents may be used to represent a large collection of information. The documents may use the same structural elements. In that case they may be linked together based on certain structural elements that are stored in an index or database. This is a common means of organizing large, highly structured collections of information such as a policies and procedures or technical manuals.

On the other hand, organization might be entirely authored through manual means using a hypertext facility. Although considerable in terms of effort, this is the only way to link collections of information that have no commonality in content or structure. Consider, for example, the development of a multimedia encyclopedia where many data types are linked by an author to help the reader traverse large quantities of dynamically related information.

But we have defined only one slice of the document, because documents do not exist for only an instant. They have a life cycle that we must also consider. During this life cycle a document may change many times: it may be annotated, it may be linked to other documents and thereby change in its organization, or it may incorporate other elements that change its structure. This requires us to consider the element of time as a fourth element of every document. This speaks directly to the issue of workflow, which is an inherent part of the document transfer and dissemination process. Next, we will discuss each of the following in more detail:

- Managing document workflow
- Managing document objects
- Managing document volatility
- Managing document sharing
- Managing document usage

Managing Document Workflow

One of the most apparent applications of workflow is in the area of EDMS. Today's organizations are flooded not only with paper, but increasingly with electronic documents. A paperless office may have sounded like nirvana some time ago, but we have come to realize that paperless does not mean problemless. The electronic document is, in many ways, more vulnerable to the ills of poor organization than its paper predecessor. In this context, workflow can be a significant asset to this new environment.

One reason for the fast acceptance of workflow is the proliferation of document-based information in desktop and LAN environments. Word processing, inexpensive document scanners and OCR, and high-capacity storage devices have all contributed to this environment. In many cases, these technologies have led to proprietary information bases for one individual or small workgroups. The result is an inefficient organization of documents with limited utility.

Nonetheless, isolated document databases are a very important source of decision support data for their users. Integrating these islands of information is a key element of workflow. It is often a difficult task, however, because the information lacks a consistent format and structure. Each group has developed its own methods and procedures for accessing and sharing information. Typically, the only method of sharing information is to transfer it in its entirety from one island of information to another. This creates redundancy, complexity, and interprocess delays. The workflow application must bring these discrete document databases together under a single umbrella and must also preserve the accessibility of the information for each workgroup. This may require heterogeneous applications and presentation formats across various platforms.

To further complicate the problem, many users believe strongly that their documents are a proprietary resource which will be compromised if integrated within a larger enterprise information base. For example, researchers and scientists at a large pharmaceutical company were reluctant to merge their analytical studies and research papers with a marketing database that would provide data for marketing and positioning of new products. In these situations the original information is thought to be of value only to the authors and historical users of the information. The sharing of information is typically accomplished through distillation of the original document into a database. Distillation is the process of extracting the key elements, or the essence, of the original and using this extraction as the basis for another group's research.

The problem presented by such a process is that other information which may be important to outsiders is no longer available. As a result the ability of the enterprise at large to benefit from its complete document information base is severely restricted. Workflow bridges this gap by providing the enterprise with full access to all documents in their original form, regardless of the document's source or the user's location.

The basic tenets of workflow analysis and the considerations that you must make for EDMS are relatively straightforward. There are three things to keep in mind about workflow applications for document management. First, do not lose sight of the fact that documents are not singular entities. Documents consist of multiple objects, each of which may be subject to distinct workflow rules and processing. Second, documents are highly volatile information types. The best example of this is the tremendous headache that most users of LAN-based systems experience when trying to control document redundancy and document versions. Finally, documents contain information objects that are not always practical for transfer over existing communications architecture. Images and audio, for example, may heavily tax networks and be almost impossible to use, in any real sense, over present-day telecommunications.

Managing Document Objects

Workflow applications attempt to orchestrate the integration and use of multiple technologies and data types. In an EDMS, workflow may be used to integrate imaging, text retrieval, and multimedia among a group of users in remote locations. The key benefits of this go beyond the simple transfer and access to common documents. Workflow provides a set of rules and a centralized tracking system by which all of the documents are managed. This ensures the integrity and consistency of distributed document processing.

The difficulty in implementing this model with many document management applications is that the workflow must recognize not only the concept of the document but also the components of the document, since each of these may in turn exist as a separate object. This level of complexity requires a strong database offering integration with the workflow system. Most likely, this should be a relational database with the ability to accommodate BLOB fields or to easily reference unstructured objects.

In addition to the importance of tracking each document object, the workflow system must be able to easily integrate with each of the applications that are creating and modifying the document objects. A document, for example, that contains a Lotus 123 spreadsheet, a Harvard Graphics bar chart, and a Word for Windows document requires a workflow system that has the ability to invoke each of these applications from a single interface. If this sounds vaguely like single point of access, it is no coincidence. Single point of access relies heavily on workflow for its effectiveness. The primary difference between traditional office automation and single point of access is the ability provided by workflow to interact with the information in a way that allows for its managed dissemination as well as its ongoing access. It is this last point which leads to the next issue of controlling document volatility through workflow.

Managing Document Volatility

Anybody who has attempted to share documents over networked environments is familiar with the problems of version control and redundancy. The problem arises out of the difficulty in working with a single document in a multiuser environment. The temptation is very strong to duplicate documents for individualized access and modification. The issue is ultimately reconciling these documents—a nearly impossible task. Workflow helps control this situation through the imposition of rules for the distribution, modification, routing, and tracking of documents.

In the first case, the document may require a series of edits by several authors. This would be coordinated by using the workflow application to manage the serialized transfer of the document to and from each author.

It may even provide information to each author as to the progress of the project. This is, however, a very simplistic workflow process.

A more complex scenario may involve simultaneous updates to different sections of the document. In this case, the workflow system would be managing a parallel set of tasks that may begin and end at separate times. This would require a set of predefined rules, or a script, to route document sections for proofreading, review, and integration of multiple information types into the final document. In both cases, the workflow system would need to work with the document as a whole as well as with its component objects.

Managing Document Sharing

By their nature, documents often contain large data types that cannot be transmitted easily over existing networks with sufficient speed to satisfy users' demands. This can have a significant impact on the nature of the workflow application and its ultimate viability. If we assume, however, that most networked environments will support some level of document transfer, the objective then becomes determining how to best utilize the local and distributed nature of each document component. Workflow now becomes intimately entwined with client/server issues, since a balance must be found between the centralized control of documents and a distributed approach where each user has local access to their documents.

It would seem that workflow in this scenario is a futile attempt since document redundancy will flourish. The fact is that workflow is mandatory in these types of distributed systems if a chaotic situation is to be avoided. Remember that one of the key components of workflow is its ability to track documents and integrate multiple applications. In this way, workflow can be used as the hub of a distributed document management system that relies on each node for a separate part of the overall application. That will allow localized control and access without loss of overall integrity.

The concept of sharing documents also leads us into the element of ownership or proprietary interest in the document. For example, documents may be secured for viewing or modification by different individuals. A document may also be modified for the purpose of presentation to a particular individual. Such is the case with a policies and procedures manual for personnel that may be accessible to a variety of employees but each employee may have limited access to certain sections or structural elements of the document. In this case the document is defined dynamically each time it is viewed based on predefined security levels. This is also referred to as a virtual document since it may exist in many forms based on certain access-time retrieval parameters.

Managing Document Usage

The final component of the document is its use. Document use is an external aspect of the document that can influence all of the components we have described thus far. Documents fall into three basic types. Depending upon your organization and your applications, you may in fact be using documents in each of the ways described. If you look closely at your organization, you will probably notice each of these types of uses in almost all document applications:

- Reference
- Transaction
- Collaboration

The reference document brings to mind the library. Libraries are very static sources of information. Books don't change as they sit on shelves. But there is great value in that source of information. There is great value in the fact that I preserved its integrity down to every single word, every single character, every single phrase. Reference applications typically use full-text retrieval. The librarian, for example, is the classic user of a keyword system or a full-text system.

Secondly, transaction documents have typically been used in the domain of image processing. In a transaction-oriented system, an application is already in place. It may be an automated information management application that handles accounting, payables, receivables, or dunning letters. All of these are transaction-based but can be easily facilitated by attaching images to the transactions. As a result imaging in transaction processing systems usually augments the existing system, causing minimal disruption. It does, however, through workflow processing, collapse the business cycle, as we pointed out in the discussion of justification.

Finally, we have collaborative or groupware document management applications. In a collaborative environment there is a great deal of interaction and change on an ongoing basis. The process may already have been in place, for example, authoring a policies and procedures manual, but document management systems facilitate the transfer of information and, as a result, provide better integration of the workgroup and the process.

It should be stressed that these aren't exclusive uses of documents. But what you will find are pockets of congestion within an organization that are characterized best by one of these uses. Unfortunately this gives the mistaken impression that one use, and therefore one technology, is suitable for a particular document management application. A reference application is ideal for text retrieval, a transaction processing application is ideal for imaging, and a collaborative application is ideal for workflow. In fact, these aren't separate applications because somewhere down the road these uses overlap and merge. As organizations within an enterprise must come together to perform a business process, so then must the document uses also come together to support that process.

The question is how quickly can you bring them together and how well your plans allow for the integration of these technologies, because transaction processing, reference applications, and collaborative applications are all different facets of the same document management problem. Part of what we must try to accomplish is to achieve an understanding of how these technologies do indeed work hand in hand. This will involve the integration of not only software and methods but also hardware and network infrastructures.

The dominant force in the integration of information systems, and certainly EDMS, is the trend toward desktop computing environments. This includes the phenomenon of proliferating microcomputers, LANs, and client/server architecture, and the role that each plays in distributed information processing and proprietary information management. By proprietary information I am referring again to the fundamental nature of document information, which unlike traditional forms of data is a highly personalized and proprietary resource.

Documents embody knowledge and hold the keys to careers and advancement. It is little wonder that individuals, workgroups, departments, and virtually every organization has its own collection of file cabinets. Whether it be for lack of trust or preservation of access, we are reluctant to give up control of documents to others. This aspect of document management has been a constant obstacle for the creation of corporate repositories and the often talked about *corporate knowledge base.*

In addition the recent move toward distributed desktop computing has empowered the individual in the organization. Traditional information systems are no longer the sole repository of the corporate knowledge base. The individual knowledge base has caused both the greatest upheaval and the greatest progress in the computing industry by personalizing information and the ability to work with vast quantities of data. The problem, of course, is that personalized computing creates disparate computing environments, redundancy, and isolated information.

Individuals are collecting enormous amounts of information for their own use. With the anticipated decreases in the cost of on-line storage these collections will continue to increase in size, by some estimates growing at the rate of 50 percent each year. The crisis this creates is orders of magnitude greater than the paper problem we've described. Documents run rampant in this scenario. Document collections are not shared; instead they are transferred from node to node, disk to disk, until a witch's brew of versions, updates, and revisions proliferates throughout the organization's desktop computing systems. Without centralization of control and coordination the EDMS loses integrity and communication grinds to a halt. Productivity actually decreases as a result of new technology investment because more time is spent administering increasingly large collections of on-line documents.

Desktop computing has indeed created an apparent Gordian knot of problems. An effective EDMS must address these issues as well if it is to provide value to an organization beyond its additional burden. Implementing an automated workflow infrastructure along with any EDMS is the primary vehicle for enabling organizations to realize this potential.

Creating a Justification for EDMS

Before we delve further into the world of the electronic document let's take a brief respite here to look at something that will act as a prelude to the use of any technology—justification of its value. If you have not already gone through the process of applying these technologies, it is difficult to impress this upon you enough: *justification is not just a technology issue.* Justification is also a business issue involving the people in the business, the organization, the structure of the organization, and the myriad cultural issues that exist in any organization.

Typically there are two sets of metrics by which to justify a new technology: the current metrics based on existing systems and costs, and the future metrics based on new opportunities and potential. The former is always easier to quantify, but the latter always provides greater payback. The problem, of course, is that there is often little evidence for the potential or opportunity of a new technology since it has not developed a wide precedent in applications or industries similar to your own. That is especially true of EDMS.

Justification Methods

- Local access to information
- Mission-critical justification
- Competitive justification
- Employee productivity
- Strategic justification
- Reengineering, entropy, and zero-based justification
- Contribution-level justification
- Collapsing the business cycle

If we consider the most often used justification metrics in each category, existing metrics and future metrics, they would at least include the following.

Local Access to Information

Instead of having to go to a warehouse where it may take 24 to 48 hours to get the documents, you can get to them immediately by using an EDMS to store the documents locally. You've reduced the time of the business cycle by doing this. That is what we will call *collapsing the business cycle.* That time period is easy to quantify as we look at the number of hours saved in retrieval time, but it is a false measure of value. Perhaps the time spent waiting was effectively used on other tasks, or it could be that the information is not required in less than 24 to 48 hours—resulting in a shorter business cycle without any tangible value. The true metric of value should not be how much faster you are able to respond to a situation with a new technology, but rather what value is added to the business process through faster response.

Another existing metric may be parallel operations. For example, while Sue is reading a document, Jack can't do anything because he has to wait until Sue is done with it; or he has to ask Sue to copy the document for him; or, worse yet, Jack goes to the file cabinet, realizes the file is gone, and doesn't know who has it. So now Jack has to spend the next two hours scouring the organization to find out who has the document. Again the benefits of this can easily be measured and translated into a shorter business cycle. But as with local access the value may not be in direct correlation to the time eliminated. As difficult as that may be to believe, it is one of the principal mistakes made by many justifiers of EDMS. Benefits that appear to be more than obvious turn out to be minimal areas of payback.

Mission-Critical Justification

The most valuable existing justification metrics result from applying EDMS to applications that fall into the category of mission-critical applications. The idea of a mission-critical application is perhaps the most often used concept for the justification of a new technology. Unfortunately, mission-critical applications are not always labeled as such. Consider your organization for a moment and try to identify a mission-critical application. Does payroll, receivable, or customer support come to mind? More than likely none of these are mission-critical applications in most organizations. Payrolls have been missed, receivable's collections are delayed, and customer's inquiries take hours or days to satisfy; yet the organization continues to thrive and grow. That is not meant to be an excuse for inefficient systems, but the fact is that most applications we consider to be mission-critical have a fair amount of forgiveness.

Consider, however, the case of major international parcel service and the catastrophic fate of an El Al 747 cargo jet that crashed near Amsterdam in 1990. The parcel service did not have off-site backup of critical electronic information prior to the crash. Although the disaster took the lives of 39 people its toll did not include the parcel service's European facility located minutes from the crash site. Data recovery had been considered cost-ineffective prior to that narrowly avoided catastrophe. After the crash it was suddenly a mission-critical application. Like the aging executive who needs a near fatal coronary before justifying the time and effort needed to modify eating habits and undertake an exercise regimen, a mission critical component is not always obvious until a similarly near fatal financial catastrophe is incurred or narrowly escaped.

Competitive Justification

The fundamental benefit of competitive advantage is fairly evident to most organizations. What is not usually talked about, however, is that organizations that are not in a position of competitive advantage are, by default, in a position of competitive disadvantage. For those industries where companies live and die by competitive rank, access to information can be their single greatest asset. In these situations justifying a strategic technology is inextricably linked with the competitive leverage it provides. Interesting situations arise from this environment since even a slight advantage can justify tremendous investments. It is no surprise that these organizations are turning to applications that in many cases are immature, isolated, and short-term solutions.

Unfortunately many smaller companies look toward these implementations as models by which to structure their own applications. Nothing could be worse.

It is, however, difficult to quantify competitive advantage. When a major Japanese automotive manufacturer came to the United States, one of its critical success factor was to be the leader in customer support. Well, how do you define being the leader in customer support? It is a very difficult measure of success. The fact is that you can look at how many times customers come back to you and buy an automobile from you as opposed to a competitor. You can identify some hard and fast milestones for measuring that after the fact. What this manufacturer did was elegant in its simplicity. It determined that when a customer calls with a complaint it must remain open until the customer decides to close the file. The customer service representative cannot close the file otherwise. It's the customer's call, so it should be the customer who decides when they are happy.

Now what does that have to do with the information systems in the background? Well, the information system has to support a very precise critical success factor: satisfy customer complaints as soon as possible. The longer the delay, the less likely the customer is to close the file a happy customer. There's no getting around that. To achieve its customer support objectives the manufacturer decided to put information systems into place that would provide immediate access to the documents required to respond to a customer complaint. For example, customer support representatives would access articles from the *Wall Street Journal* or major automotive magazines about a particular article that the customer may have read, and as a result called about. They would bring up internal comments from their analysts and researchers or images of automobiles and component diagrams. This was all part of the document management system intended to support the customer support analyst as the mission-critical component of this manufacturer's competitive advantage.

Employee Productivity

This leads to an interesting aspect of customer support and one which is impossible to overcome without an EDMS. The biggest hurdle in developing an effective customer support organization, using traditional means, is that people need time to become accustomed to the systems they are supporting. Every customer support department has groups of experts that are treasured by any customer; then there are the analysts who are fairly knowledgeable, and finally the trainees—whom no one wants to get on the phone. After all, no customer wants to end up training the customer support representative when they are in the middle of a crisis.

With a well-designed customer support system, the objective is that everyone in the support group should be as good as the best, because they all have access to the same information. But that is rarely if ever the case because traditional training and experience-building measures rely on time as the yardstick of competency. In addition, the knowledge gained by the few exceptional individuals within the group is often closely guarded by those same few. It is, after all, this quality which makes them unique and most valuable.

An EDMS can, if correctly used, bridge if not altogether eliminate the knowledge gaps in a group of individuals by making information readily available to everyone. This goes beyond simply referencing and retrieving the experiences and scenarios of others. More importantly it creates an accelerated learning environment that provides greater opportunity for those who rely on the knowledge base to develop their own experience sets. That produces a highly qualified expert in only a fraction of the time normally required. The long-term impact is a decrease in the number of analysts needed to support a given volume of inquiries. It also leads to the elimination of the expert bottleneck that results when a few experts exist to answer a large volume of questions filtered through a long line of less-experienced analysts.

This was demonstrated in the *U.S. Patent and Trade Mark Office* (PTO). When the PTO brought in an imaging system, the biggest single benefit it produced was not necessarily the most obvious. It was not parallel access to information, which is fairly important when you're dealing with millions of paper-based documents. It was not local access to information, which is also important when dealing with warehouses of documents. It was not the savings of floor space and file cabinets that were no longer needed for storage. It was not even speed of access, which is hardly insignificant when you consider that a patent takes about a year to get through the PTO system. The biggest single benefit, according to the commissioner of the PTO, was that new patent examiners came up to speed faster. It didn't take 12 years to develop a tenured patent examiner. New examiners could be up and running in a few months.

Strategic Justification

Although the justification of a strategic technology such as document management must start with a careful assessment of an organization's business objectives, it should not look only at the current business practices and models of the organization. Going further to evaluate the future potential that document management may bring to the organization involves identifying the specific means by which an organization measures its effectiveness and, as a result, gauges a technology's contribution to its success. These are typically referred to as the *critical success factors* (CSFs) of a business. For example, a bank's CSFs may be based on effective risk avoidance or outstanding customer service. A pharmaceutical company would probably emphasize time to market in defining its critical success factors. Of course, the specific CSFs used will vary from one organization to the next, and may even differ within an organization from one profit center to another.

Although there may be a combination of forces at play, one set of CSFs will always be predominant at an enterprise level. This is the level that should be focused on since, once identified, this characteristic of the business becomes the driving force in the organization's justification of any new strategic technology. Trying to assess the costs of any strategic investment apart from the CSFs is similar to choosing inexpensive cast iron instead of titanium for the wings of a fighter jet. That may well minimize costs, but it does not serve the primary objective of the plane, which is to fly.

CSFs focus on results well ahead of defining any specific underlying tasks or technologies which increase or reduce costs. In fact, assigning costs to these tasks in a future context makes little sense, since they will be significantly altered as a result of the reengineering process inherent in the implementation of a document management system. Evaluators who do not heed these changes run the risk of implementing point solutions. Although point solutions may provide a short-term benefit within a well-defined niche, they do not typify the strategic nature of most enterprise applications. By considering the CSFs and aligning the document management solution with them you are able to provide for the foundation of much more dramatic change and, in turn, much higher payback. That is the basic premise behind cost-justifying document management based on a reengineering of the organization's information systems.

Reengineering, Entropy, and Zero-Based Justification

The life cycle of any information system is limited by unavoidable factors. Changing demands and usage patterns slowly wear away at the utility of an information system until the process of entropy takes a firm hold. At this stage, the information system is no longer capable of satisfying users' needs and requirements. In fact it becomes an impediment to effective information retrieval. The cost of entropy, in the context of information systems, can be equated to that of using outdated machines and tools for a manufacturing process. The tools may be maintained meticulously but they are nonetheless inadequate for the demands of modern manufacturing tasks.

To follow the analogy, there is only so much that can be done to adapt the old machinery to a new process. After a certain point the entropy effect cannot be reversed. The only course of action is reengineering. Reengineering does more than replace the old machines with new ones. Reengineering also redefines the process. This is why automobile manufacturers are no longer stocking vast warehouses of parts for automobile assembly. Robotics have made it possible to project production requirements precisely, thus allowing for *just-in-time* (JIT) delivery of parts and many other new manufacturing methods never before considered.

In a similar way, reengineering of an information system through the application of document management technology ultimately requires the restructuring of the information processing cycle from collection to dissemination. The advantage extends far beyond the increased functionality of the new system to the elimination of time-consuming and ineffective methods of information management. The effect of this on cost justification is a shift in focus from evaluating technology based on the cost of old technology versus the cost of new technology to an emphasis on zero-based cost justification. In a zero-based model there are two rules: there is no assumed cost of doing business; and every expense may be partially or entirely allocated to an existing information system. The difference this makes is to equalize existing and new technologies.

Contribution-Level Justification

Although it is impossible to predict all of the costs of the existing system and all of the benefits of the automated system, there are some basic costs which are not usually considered in a classic analysis but are crucial to justifying document management solutions. These costs may not be a direct function of operating and maintaining existing systems, but they are at least as important to consider. However, the traditional concept of opportunity cost does not fit well into a justification, since it gives the impression of excessive optimism on the part of the justifier.

A more appropriate concept is that of *contribution level.* This subtle but important modification of the cost justification can make a world of difference. An existing information system may, for example, result in the creation of excess inventory, which will lead ultimately to spoilage. The spoilage would normally be considered a cost of doing business; instead, the spoilage should be a cost of the existing information system. The new technology is justified partly on this basis, its contribution level. Think of contribution level as the information system's proactive avoidance of costs or the proactive support of revenue generation. In this light the information system plays a role in every line of an organization's income statement.

Such was the case at First Brands, a major manufacturer of consumer products such as STP oil additives. First Brands had a problem with the production of excess inventory during the creation of packaging for its products. Due to the petrochemical origin of many of First Brands' products the regulations for packaging and labeling were very stringent and subject to frequent change. That required new artwork and labeling instructions to be sent from corporate to the manufacturing facility in time to alter the packaging for a particular product. The problem, if it can be called that, was the tremendous demand for First Brands' products. Production of packaging runs at a frenetic pace to keep up with consumer demand. As a result the new packaging instructions sometimes arrived at the packaging production line too late to stem the waste of many outdated, mislabeled containers. In a healthy and thriving business this is simply attributed to the cost of doing business. At First Brands, although manufacturing was running at peak efficiency and corporate was responding to packaging changes instantly, the information system, a combination of overnight couriers and facsimile, was not adequate to handle the transfer of information from one department to the other.

At the same time the cost of business attitude stood firmly in the way of justifying an EDMS. A subtle shift in the way of thinking paved the road for a justification. The cost of wasted packaging was attributed to the existing information system. Suddenly the payback analysis leaned dramatically toward a new information system, which would pay for itself quickly, considering the high carrying cost of the current system.

As easy as that example may be to understand in retrospect, far too many organizations have buried the cost of their ineffective information systems in the costs of doing business. In a future context nothing will change if this archaic attitude toward costs does not also change.

Collapsing the Business Cycle

Finally, one of the most dramatic effects of document management and one of the most powerful cost justifications is the shortening of business cycles. This is a function of the increase in available facilities for communicating and accessing information in a timely fashion. The result is familiar to those of us who have made the transition to E-mail from paper-based memos and interoffice envelopes. Once the obstacle of information transfer is removed, every aspect of the organization benefits from the reduction of turnaround time that is inherent to paper-based systems.

A further advantage, however, is not as easy to perceive. Most new users of a document management application soon realize that the time required for tasks which were not feasible and impractical can be reduced dramatically. This allows for research and decision support that would not have even been considered before automation. The impact that collapsing the business cycle has on the bottom line can be measured easily in a number of ways: faster and higher-quality decision making, shortened research and development time, and reduced sales cycles. Each situation will present its own set of benefits which can be quantified and factored into the justification. We will look at this in much more detail in Chap. 3.

3
Workflow

Productivity Standstill

A disturbing element is common to all of the technology advances in office computing over the last two decades. Despite monumental investment in information systems and technology literacy, the white-collar work force has experienced a virtual standstill in productivity over the same time period.[1] This fact seems to fly in the face of constantly increasing output. But that increase has not been without significant cost. Output has increased in proportion to capital expenditures. The result has been a net productivity gain in the white-collar work force of less than 1 percent over 20 years. During the same period, productivity increased on the factory floor at double-digit rates. It is obvious that information technology has failed to deliver on the promise of productivity in the office environment.

The Role of Workflow

Workflow applies many of the same concepts and benefits of factory automation to the process of work management in an office environment. Workflow provides a tool set for the proactive analysis, compression, and automation of information-based tasks and activities. The basic premise of workflow is that an office environment is an information factory, or more specifically a document factory. The document, which can exist in a range of formats from paper to electronic form, is the basic raw material of every office task. The connection of office tasks creates a value chain that spans internal and external task boundaries. In this architecture, workflow attempts to streamline the

[1]This claim is often quoted. For detail the reader is encouraged to review many of the writings of Lester Thurow. A more optimistic viewpoint may be found in the works of Paul Stuassman.

components of the document factory by eliminating unnecessary tasks and thereby saving time, effort, and costs associated with the performance of those tasks and automation of the remaining tasks.

Evolving Office Systems Require Workflow

Only recently have office systems begun to evolve from isolated personal productivity tools into networked resources for document transfer across organizations. This trend toward connectivity among office workers will undoubtedly prove to be one of the most significant events in the history of white-collar productivity to date. There are those who would say that this is the natural evolution of interconnected desktop computing, but much more is involved in workflow. These extended coalitions of office workers have created a new work environment which requires new tools for the coordination of their activities and document-based interactions. The foundation of these tools is workflow technology.

The Document Factory

A revolution occurred in factories as assembly lines consisting of discrete tasks gave way to tightly coordinated parallel processes of manufacturing, inventory management, and retooling for mass customization. Workflow will similarly revolutionize the "document factory" as it provides for a streamlining and integration of information-based activities at levels of integrity and reliability never before possible. These last two points are significant in an era when organizations are downsizing to smaller work forces and facing greater competitive pressure.

It should be noted however, that the factory analogy has its shortcomings when applied in a literal sense to an office environment. Factory environments can be highly structured and disciplined. Tasks may occur with relentless precision and consistency. Errors are reduced to zero defects and hair-width tolerance. In a factory, this type of environment can be maintained for long periods of time without flaw. In an office environment this same approach will soon result in reduced productivity and lost opportunity as knowledge workers lose the ability to structure their own time and creative process. An office environment may thrive on a measure of creative chaos. Periods of high productivity are often preceded and followed by what would be considered idle time on the assembly line. If managers and developers of workflow systems do not recognize these fundamental differences, they will not only frustrate users, but will also undermine the potential for increased productivity.

The factory productivity phenomenon can be replicated in the office by applying some of the same principles that were used on the factory floor, but with a slight modification. Factory automation disciplines such as *total quality management* (TQM), *electronic data interchange* (EDI), *just-in-time* (JIT) methods, and *time-based analysis* are inherent in workflow technology.

Redefining the Information Asset

One of the most important aspects of the advent of workflow is the impact it has had on the concept of information as an organizational asset. The premise of information as a tangible asset that must be preserved and valued does not change. What does change is the concept of information. Traditionally, the information represented the data and documents that were used in the process of supporting decision-making activity.

With workflow the information becomes twofold: the data and the process. Workflow enables an organization to capture the process itself, including the decisions, the variables of each processing rule, and ultimately the roles of each individual. Before chords of fear begin to ring for those readers who see this as yet one more way to downsize the work force, stop and think about the flip side of the downsizing equation. Fewer people are tasked with the increasing burden of performing more and more tasks.

Without a means by which to validate, compress, and automate work roles, downsizing has no long-term benefit. Workflow not only automates the process but can actually improve the standard of work environments by minimizing miscommunications among heavily burdened workgroups.

TQM and Workflow

In the context of the office factory, workflow realigns the overall approach to managing information with a process known as Total Quality Management (TQM). TQM is a broad-based model for evaluating and redeploying every aspect of the organization toward the final goal of increased service and product quality.[2] Information moves quickly and directly to the individuals closest to the process. These individuals are presumed to be best able to determine what problems may exist and what measures must be taken to improve quality.

[2] *Putting Total Quality Management to Work: What TQM Means...*, Sashkin, Marshall, and Kiser, Berrett-Koehler, 1993.

Workflow facilitates TQM by providing tools that bridge potential gaps between layers of management. In fact, the approach is in some cases so effective that the ultimate result is a flattening or restructuring of the organization to eliminate layers of inefficiency. In these new organizational models, workers are empowered with information and the authority to actively contribute to the quality process. The TQM environment generates a much higher need for ongoing communication and coordination at a peer-to-peer level since the traditional management hierarchy no longer provides top-down coordination. Tools such as workflow ensure the integration of these tasks through automation of the information transfer process. The benefit is improved communication and better responsiveness. The result is a more competitive model for a quickly changing marketplace than that provided in an extensive and tedious management hierarchy.

Workflow in the Extended Enterprise

Recent advances in technologies such as EDI have brought about a new model for describing the structure of an organization's information systems. This new model, called the *extended enterprise,* has far-reaching implications on workflow technology since it causes the workflow application to extend beyond internal data. In the extended enterprise, an organization's information systems are no longer bound to internally generated data and documents. Instead they are redefined to include all documents and information that apply to a particular task or process. This could extend well beyond the enterprise into other organizations.

A prime example of the extended enterprise is the expanding use of EDI where business information is shared among trading partners and across organizational borders to expedite transactions. Trading partners share certain common information with access rights that allow direct access to each other's information. For example, through the EDI model, an order placed by Ford for tires supplied by Goodyear would give Ford direct access to Goodyear's inventory information. Ford may be able to initiate a manufacturing process at Goodyear, which corresponds to Ford's manufacturing schedule, based on production times and schedules at the Goodyear factory.

The challenge posed by this model is the establishment of parameters and rules for the integration of the two business partners' manufacturing activities. This is where workflow becomes a major factor. The workflow model would provide the tools for defining these parameters, the integration of discrete applications, and the rules of any extended enterprise transaction. The workflow model would work just as well with a system that integrates office functions such as sales and marketing as it does with the shop floor functions.

Time-Based Workflow Analysis

One of the most pronounced benefits of workflow is the elimination of idle transfer times. In every office environment two elements affect the overall efficiency of a business cycle: task time and transfer time. Although either one can slow down a process, transfer time is usually the worst and best-hidden villain in an organizational setting. Transfer time is most typically ignored during an analysis of existing procedures because focus tends to be placed on the people and the tasks and not on the time that passes between the completion of individual tasks.[3] Workflow addresses both task and transfer time and can effectively collapse the idle period between tasks. Although the result may appear to be a function of using electronic information as opposed to paper-based communications, it is far more involved.

Electronic information can be accompanied by many of the same problems that plague paper-based systems. For example, a large East Coast insurance company was surprised to find its pilot imaging system did not expedite claims processing. On closer inspection, it turned out the claims processors had not adapted to a new workflow model. Rather than work on claims individually as they were received, the processors would wait until a week's worth had accumulated before processing a single one, just as they had done with the manual system. Had the company performed a Time-Based Workflow Analysis, they would have realized that its existing processes were inefficient. Processes could have been redesigned to take full advantage of the capabilities of workflow and imaging technology.

The Workflow Environment

It can rightfully be said that workflow brings little to an organization that does not already exist in office automation technologies. What workflow does is to consolidate existing office automation and document management technologies under a single environment. It is, effectively, a complex distributed data manager. Workflow often represents the critical link between technology and people. When it is missing, the whole system suffers. The conversion to electronic documents

[3]Time-Based Analysis™ methodology, *The Workflow Imperative*, Koulopoulos, Van Nostrand, 1995.

in itself does not add the intelligence and control necessary to achieve the benefits of process automation. Workflow provides this shell for electronic documents. It is this uniting quality of workflow that is most often misunderstood.

The idea behind workflow technology is to create a single environment to manage the complexity of multiple office automation environments. As software and data have migrated from individualized solutions with dedicated functionality, to integrated solutions, and then to groupware solutions, workflow has evolved as a metaphor for the coordination of multiple workgroups using multiple technologies. This parallels the point-of-access paradigm we introduced in Chap. 1.

Initially workflow enters an organization as a two-dimensional solution, handling multiple technologies used by multiple individuals within a workgroup. In time, the system is expanded for use by many different workgroups, transforming it into a three-dimensional model shown in Fig. 3-1. For many organizations, this may be one of very few opportunities to share electronic information and perform electronic transactions across workgroup boundaries in a single cohesive environment. The complexity of the workflow model grows quickly at this stage, as user requirements and the diversity of information increases. This three-dimensional metaphor is important to understand in planning for workflow. Organizations must carefully select workflow products with a range of tools that supports a variety of office automation technologies and allows for the development and orchestration of this diverse environment.

Creating Workflow Environments

Workflow applications are written with a variety of tools from a scripting language (similar to a 4GL) to an object-oriented CASE method utilizing icon-based GUIs. Many older workflow scripting languages were complex and required a fair amount of training and experience with the tools and command sets. Early versions of workflow automation (prior to 1991) are examples of this type of product. Most products now abstract the scripts into a series of objects which can be modified with forms-based entry or graphical models. This may limit the overall flexibility in highly customized situations, but it is much simpler for the development of immediate solutions by less-than-expert developers.

We should note that many business processes are very complex and even with easy-to-use tools for workflow development, end users may not be able to develop enterprise-wide workflow applications. This is one of the misconceptions about many new workflow tools. Although the specific tools required to create these environments are easy to use,

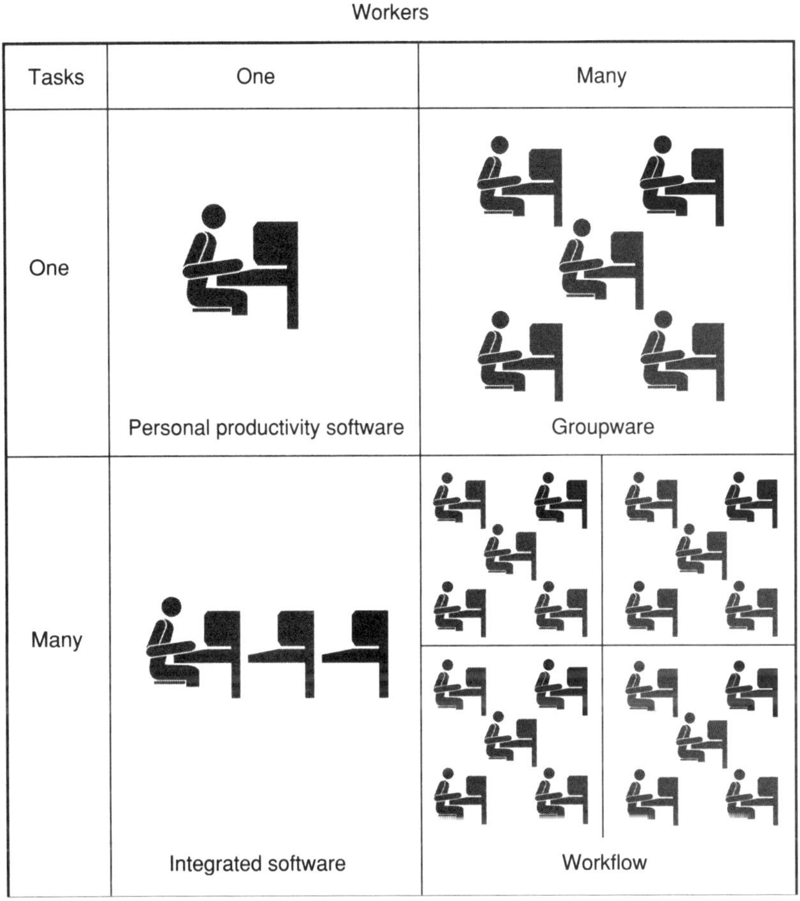

Workers

Tasks	One	Many
One	Personal productivity software	Groupware
Many	Integrated software	Workflow

Figure 3-1 Workflow operates at the highest level of automation. The development of these software categories has paralleled the development of hardware technology: (1) Personal productivity software became a significant market with the advent of the personal computer. (2) More powerful personal computers permitted integrated, multifunction software. (3) PC networking enabled groupware. (4) Workflow software was developed, based on a client/server architecture that traverses multiple organizations within the enterprise.

the complexity of designing, implementing, and maintaining enterprise workflow applications still requires the involvement of IS.

Reengineering and Workflow

Reengineering is often thought of as the radical process of decimating and then rebuilding information systems. In many cases, workflow is part of the reengineering process. Workflow does not, however, require radical change in the organization. In fact, most workflow applications are facilitated by the use of an incremental or pilot implementation of the technology. This provides a tremendous amount of experience to the implementing organization. It also shows immediate results, which provide for a leverage point in justifying future enterprise applications. This is especially true of workgroup solutions. As the sponsorship and scope of the workflow application rise within the organization, this will shift dramatically. At higher levels the methodology, planning, and simulation of workflow are essential to delivering solutions in a reasonable time frame. Between these two extremes is a break even threshold below which an educational approach is the best option and above which emphasis should be placed on the methodology.

Identifying this threshold is essential to avoiding the pitfall of investing too much or too little time in the analysis phase of defining a workflow application. Generally, the number and size of the workgroups to be included in the workflow systems are good indications of the analysis required. Single workgroups of less than 25 workers do not typically require more than a basic assessment of the work environment and a brief exposure to the pilot application to determine their specific requirements. Workgroups of between 25 to 50 individuals may benefit from a Time-Based Workflow Analysis[4] and a basic pilot application, again primarily as a testing ground for quick follow-on applications. Larger workgroup and multiple workgroup solutions that go beyond 1000 users require up to 90 percent of the overall investment to be made in the area of business analysis and workflow modeling. In this last instance, formalized methodologies, such as those presented in the implementation section of this report, or public domain methods such as IDEF, are absolute mandates to effective workflow implementation. At this level workflow and reengineering become closely related.

There Is More Than One Way to Reengineer

Reengineering is not a single-dimensional model. There are several distinct models of reengineering. Among the most often encountered are the entropy model, the life-cycle model, and the goal-oriented model.

The three methods of reengineering:

[4]Time-Based Workflow Analysis™, *The Workflow Imperative*, Koulopoulos, Van Nostrand, 1995.

- The crisis model
- The life-cycle model
- The goal-oriented model

The Crisis Model. The crisis model of reengineering results from internal or external pressures in the organization that mandate a decision to alter an existing information process. This is often used as an opportunity to take advantage of the change for greater benefit than the problem at hand. Typically the issue of sponsorship is irrelevant since the effort must be expended, in any event, to change the system. It is unlikely, however, that a methodology will be applied to this type of reengineering. For example, a manufacturer of government-regulated products may simply not be able to meet OSHA regulatory requirements utilizing its existing information systems, thus mandating a new approach if it is to continue doing business under OSHA regulations.

The Life-Cycle Model. The life-cycle model of reengineering is an ongoing method of reengineering that recognizes the benefits of constant refinement of existing systems. In these environments, users and developers are closely linked through ongoing dialogue on the benefits and shortcomings of the systems. Management is constantly evaluating the cost and benefit of existing versus new solutions. There is a track record of new investment that demonstrates the payback of new technologies that bring improved methods, but more importantly there is also a progressive attitude toward new approaches and methods—with or without regard to technology.

The Goal-Oriented Model. Sponsorship is an inherent issue in goal-oriented reengineering. Goal-oriented reengineering is undertaken with well-defined objectives in mind. It is a deliberate attempt to bring an existing process or activity in line with business objectives. For example, a service company may decide to reengineer the customer billing process to speed up the invoicing of customers and improve cash flow. Goal-oriented reengineering such as this requires a sponsor with the ability to drive the application through the organization—in many cases without an adequate cost justification.

Human Factors: Corporate Culture and Workflow

The largest single obstacle faced by organizations planning to implement workflow is that of existing organizational culture. Unfortunately, most information professionals do not easily accept the fact that the real issues are not technology-based but rather are mired in human factors and organizational issues. The reality is that culture accounts for 67 percent of the obstacles identified by evaluators and users of workflow. If these issues are not addressed, the success of workflow is compromised to a substantial degree.

The cultural effects workflow can have on the organization are numerous: the flattening of the organization's management structure, the imposition of a new level of control on the office workers, increased availability of information to the point of overload, unnecessary structure, and diminished proprietary interest in processes. Although each of these can be turned around with some forethought and planning, they can each be equally devastating to the long-term viability of the workflow systems if not given careful consideration at the outset.

Flatter Organizations

Flattening of the organization's management structure is a mixed blessing as we have already seen in our discussion about TQM. It is not, however, an inevitable result of implementing workflow technology. In the climate of downsizing that is prevalent today, organizations are already running lean. The problem for them is not in eliminating idle capacity and flattening the organization, but rather in coping with this model after it has been imposed by existing business conditions. Workflow is a tool for adapting to the already changing structure of the organization. It enables organizations to restructure more effectively and preserve stable market share and business volume with reduced staffing.

Lost Loopholes

This flattening of the organization has other implications as well. It imposes a new level of control on office workers. Loopholes and hidden inefficiencies become evident as existing processes are analyzed and eliminated through the development of automated workflow rules. These rules are enforced and carefully monitored for process efficiency. The irony is that some level of flexibility and an occasional loophole allows for the introduction of creativity into the process. This can be a productive element of office systems that support empowered user models such as TQM. If the workflow system is arbitrarily rigid and tightly structured it will inhibit the ability of workers to participate and contribute to dynamic and spontaneous communications.

More Information Than Ever

In addition, these same workers will face enormously increased amounts of information, to the point of overload. Since the natural ebb and flow of work will be regulated and transfer times minimized or eliminated entirely, the worker's queue will be constantly demanding attention. The more efficient the business, the greater the likelihood that the user will be the bottleneck. It is the classic man-to-machine interface problem. The workflow system must be regulated and coordinated with the capabilities and the education of the worker in mind. That may mean increasing the worker's understanding of the process through education or it may mean tempering the zeal for productivity to favor models that evolve with the workers.

The Proprietary Nature of Knowledge

Finally, if all else is ignored, the worker will lose what is perhaps the most valuable aspect of office information, the proprietary interest in the process or the information. Office tasks and documents are closely held by workers as proprietary material, evidenced by the preponderance of filing cabinets in offices and the size of disk drives on user's desktop machines. Information and its ownership means control and security. Careers and livelihoods are determined by specialization of knowledge and availability of information. Take this away from workers and they lose interest in the product, service, or task at hand. It is no different on the assembly line. Workers enjoy a sense of pride and benefit from their ownership interest in their knowledge. When that knowledge is embodied in paper or electronic information, it becomes a linchpin of their job security.

Although each of these obstacles can be overcome with some forethought and the enactment of the measures described, they are a natural part of the evolution of a workflow application. Some of these can be anticipated; some cannot. An approach that stresses education, an emphasis on the business, an appreciation of the cultural issues involved, and an incremental approach to implementation will provide the greatest likelihood of success and user acceptance of workflow.

Technology Components of Workflow

Before we can examine the technology components of workflow we need to understand its evolution. Workflow evolved as a means by which to take advantage of the networking infrastructure, integrated presentation environments, workgroup management, and office communication tools under a single umbrella. Placing workflow in this context helps one to appreciate the soundness of workflow technology and its role in today's information systems.

Workflow has evolved out of several established technologies including E-mail, project management, database, object-oriented programming, and CASE. With more than 70 workflow products in the market, there are a variety of approaches combining different aspects from each of these technologies.

The different workflow approaches and solutions are similar, however, in one fundamental manner—they all attempt to solve the problem of fragmented and isolated information. Only recently has this been possible to consider. If we reconsider our paradigm model of information technology evolution, you will recall that there are three distinct phases of information systems that have led to this present-day model of shared information and the need for workflow. To recap the earlier discussion:

> First paradigm technologies (circa 1950s to 1960s) were centered on the computer program that drove the information processing in a highly isolated model—the von Neumann model of computing.
>
> Second paradigm technologies (circa 1970s to 1980s) shifted the focus to the data models in a highly centralized environment. These were typified by mainframe and large database applications.
>
> As we entered the third paradigm in the 1990s emphasis again shifted, but this time away from the technology toward business processes and tasks of which they are composed. The third paradigm brings full-range computing into the personal domain of individual workers at all levels of the organization. For the first time it is possible to create and easily propagate isolated computing environments throughout an enterprise.

It was the advent of third paradigm information systems that resulted in true "islands" of information with virtually no connection to the rest of the organization.

As these independent systems have proliferated, they have become vast warehouses of information. The information they contain has become the foundation for individualized empowerment and advantage. Systems can be customized to meet the needs of each user, in many cases, directly by the user. From the user's standpoint, diversity is a strength.

But there is a downside to this success and rapid proliferation. Incompatibility has flourished among applications and data repositories, across vertical and horizontal lines of the organization. The response is ironic: more paper than ever before is being generated by computer-based systems. Perhaps the greatest beneficiaries of all this are the suppliers of printing hardware. Seven hundred million pages of computer output are generated each day and 70 percent of that paper is used for data entry into other computer systems—a testimonial to the inability of different computer environments to communicate. The information archipelago has arrived and the bridges from island to island are paved with paper!

We refer to workflow as a third paradigm information technology due to its emphasis on task-centered models and efficiencies. Workflow has evolved from this environment as a result of the migration toward distributed workgroup computing and higher-level languages for the definition of process automation. It is not so much a radical new technology as it is a logical next step in the use of computing to integrate knowledge-based tasks and activities. Its primary benefit is joining islands of automation, which use discrete tool sets, into enterprise information systems. In this sense, workflow is an operating environment that provides a context for working with other technologies.

Understanding the nature of this third paradigm and the evolution of workflow, we can look at the technology components of workflow, understanding that workflow is not a singular application but rather the orchestration of many applications and data types.

The Process and Data Models

Workflow applications use seven distinct structural components to define the organization of information management activities.

The Process Model

The process model, the highest layer, consists of a series of tasks and rules which must be defined for each process that takes place. For example, an accounts payable process is made up of several tasks such as the receipt of the invoice, review and approval, reconciliation with the receiving report, etc. Specific tasks have rules associated with them. For example, invoices over a certain dollar amount will require a higher level of approval. Each task and its associated rules must be defined to the workflow system. We will explain later how the rules layer of this model is broken out further into a series of attributes that define the way in which documents interact with the process and folder layers.

Cases

The next layer is the case, which is an individual occurrence of the process model. The processing of a single invoice for payment would be considered a case in the accounts payable process model. Each time the workflow procedure is invoked a new case is created.

Folders

The third layer is the folder which contains a logical group of documents. The folder may contain any combination of data types including text, image, and data from multiple sources. A folder for an accounts payable application may contain the purchase order, receiving report, and financial data from the general accounting system.

Rules

The fourth layer is the rules associated directly with the documents or data. These rules define the specific processing activities that are involved in routing workflow data and working within specific applications. For example, a particular contract document may be associated with a schedule of approvals. The approvals are governed by rules that relate to the amount of the contract. In addition, an approval may in turn launch an application that updates a spreadsheet cell.

Data

At the lowest level is the data itself. This is represented by a single document or data type which is related to a folder and case. It should be noted, however, that the document is almost always associated with an application that is initiated in order to view and work with the data. The processing of an accounts payable invoice may require viewing of the purchase order and relevant financial reports. The workflow system must be able to launch the appropriate programs to enable use of the information. The ability to invoke and integrate the presentation of these documents is a key aspect of workflow.

Roles

The definition of a specific workflow requires the establishment of roles for all participants. Each participant has established roles which must be defined as part of the workflow definition.

Each participant is described in terms of location, job function, supervisor, and security level. An employee may be an individual participant as well as part of a workgroup. Workgroups can consist of a group of individuals working on a project, a group of the individuals in a department, or a group of individuals that share the same job function. One individual may belong to multiple workgroups at one time, depending upon the type of application.

Role definition enables the workflow system to distribute tasks to the appropriate individuals for completion. It also allows for work load balancing and distribution to the appropriate employee based on skill level.

Routing

Workflow routing rules control how documents move from point to point in the workflow. The three types of workflow routing are sequential, parallel, and dynamic or conditional routing.

Sequential routing follows a linear path from one task to another. It is clearly defined with little variation. One task must be completed before the work is routed to the next point.

Parallel routing enables multiple tasks to occur at the same time. In an accounts payable application, for example, a document may require the approval of three different individuals. Since the three approvals are independent of one another, they can occur simultaneously. Eventually all three are brought together at a rendezvous point. They are held there until all approvals are obtained before initiating the next task.

Conditional routing is determined by conditions which occur dynamically in the process. The system must be able to determine the appropriate route based upon the information that is received along the way. For example, an accounts payable invoice over $10,000 requires an approval of a vice president in addition to the department director. Upon receipt, the workflow system that incorporates dynamic routing would automatically route all invoices over $10,000 to the appropriate vice president for approval.

Workflow Tools

Workflow Scripting Languages

Workflow scripting languages are used as a method of writing code to define processes, rules, and operations for the workflow application. Object-oriented and 4GL languages are the most common scripting tools used in designing workflow scripts. These tools are used mainly by professional developers to design and generate code. Although easier to use than older scripting languages, they are not easily modified by end users. Workflow scripting languages are most commonly used in transaction-based workflow applications.

Graphical Workflow Editors

Graphical workflow editors are used by both professional designers and end users to develop workflow applications through the use of graphical tools which guide the user through the process by posing a series of questions in a menu or by creating on-screen flowcharts that depict the workflow process. The icon-based approach is often easier to use than workflow scripting languages since it relies on familiar metaphors for windowing environments, such as point and shoot, click and drag, and radio buttons. Workflow applications developed using graphical editors are easier to modify and so provide the end user with greater flexibility in the development and modification of workflow applications, without the reliance on professional developers. These advanced interfaces may, however, require additional programming in order to adequately integrate with other office automation technologies. Graphical editors are used in all workflow types but are most common in ad hoc applications where there is a high reliance on the end user for development of individualized workflows.

Workflow APIs

Workflow *applications program interfaces* (APIs) are provided with most workflow products. They are essential for the integration of workflow with other business applications. The critical aspect of the API goes beyond simply sharing information, however. The API must be able to ensure the integrity of the information that is accessed and transmitted. A workflow application intended to integrate a spreadsheet tool and a customized management report must be able to share data. Changes made to the spreadsheet that have an effect on the management report must be reflected in the report. Conversely, changes made in the management report that affect the spreadsheet must be reflected there as well. This type of dynamic linkage is an important aspect of establishing the workflow environment since common data elements are shared by a variety of applications across the enterprise. Workflow tools that manage this data can have a dramatic impact on the workflow system's ability to do this.

Workflow Simulation Tools

Workflow simulation tools allow designers to test workflow processes and to identify problems with the process before the workflow application is implemented. At the present time very few workflow systems provide these facilities as part of the workflow product. Many developers rely on third-party vendors that have software simulation tools suitable for business modeling. Public domain tools such as *integrated definition* (IDEF), which is used extensively by government, can also be used to assist in this process.

Databases

In addition to workflow development tools, workflow products include databases that are used for the storage of routing instructions, procedures, and rules. Databases are also used to track the status of workflow processes and to maintain a historical audit trail of each transaction. It is here that many workflow products fall short. The database and the concept of a data dictionary are not yet fully developed in most workflow systems. The data dictionary provides a singular definition of the attributes associated with each data item or document, but the variety of sources of documents and data elements in a workflow system is so robust that it may be next to impossible to impose a singular data dictionary.

Most workflow products use either standard third-party database products or incorporate their own proprietary databases. Typically, integration with external databases is provided by using existing APIs that call external databases such as ORACLE, Sybase, or other standard DBMS packages. In some cases, vendors provide their own proprietary database scheme, which cannot be accessed as a table or connected to an external DBMS. These approaches can hinder the implementation of workflow as a user environment and result in the restrictive use of workflow in isolated workgroups or simply as a document management tool for transferring and tracking. This is not a death knell for the technology since it is still advantageous to begin joining workgroups together in advance of enterprise applications. But it does cause problems when attempting to create larger systems that are intended to join all elements of an enterprise's office automation technologies.

Recent workflow products use object-oriented databases. Object-oriented technology is ideally suited for workflow software because of its ability to represent complex information without compromising flexibility. Each object in an object-oriented database can be defined and maintained independently of one another and each is defined through a series of procedures (methods) and data elements (variables).

Unlike the more structured relational databases, object-oriented database objects interact in a multitude of ways. They simulate real-world activities and can be easily manipulated and modified. Object-oriented functions such as inheritance and object classes enhance workflow applications by avoiding the repetitive creation of objects and preserving the object's attributes throughout all process models. The concept of object classes allows for the creation of a structured hierarchy of workflow objects into super classes and subclasses. For example, purchase orders may be a subclass of the procurement process model while at the same time being a superclass for order processing. Encapsulation masks all of this complexity within the concept of a single object or icon that can be bundled with any combination of other objects for incorporation into a new workflow.

Workflow technology should be tightly integrated with enterprise-wide database technology. The issue of referential integrity among all of the documents and data sets must also be considered. This ensures that changes made within one application are reflected in all other applications also using the data. Administration should also be considered since it will be an important part of constructing and maintaining this new environment. But the database integration has other benefits as well. If a multimedia, or BLOB, database is used it will provide ready access to a broad variety of data types, including documents of all forms.

The Time-Based, Event-Driven Model

There are three basic components of every workflow environment: events, time, and objects. Workflow technology is founded on the principle of coordinating events within an optimal time frame by defining data objects that recognize specific rules. Another way to look at this is that events and time are the basic building blocks of objects of a workflow system. Creating these objects, however, requires a set of tools for defining each one and the relationships among them. Here, too, the availability of actual product functionality and the requirements of workflow developers may differ considerably.

Current workflow implementations use procedural methods for accomplishing the definition of events and duration. For example, the product may provide an object-oriented definition language for creating a task. The task is represented by an icon that in turn has been defined as a set of rules for the task. These may be the approvals required for a particular document, or the integrity controls for references to external data sources. The object will also contain information about the time line that the document should travel, such as that allowed for approval before the document is rerouted to another individual.

An effective workflow tool set should provide support for object creation, management, modification, and linkage in much the same way as object-oriented programming tools. Functions such as inheritance, object classes, and encapsulation should also be supported by the workflow system in order to avoid repetitive creation of objects and to preserve the object's attributes throughout all process models in which it is used.

Each object has six of these associated attributes that reflect activities inherent to every information-based activity or business cycle. These six attributes are initiation, notification, iteration, completion, dependence, and duration. Each attribute exists in any transfer of information that takes part in a value-added process, a single task, or a process in a workflow cycle, also called an *event*.

Initiation (also called a trigger). An event that begins a workflow process. This may be as simple as an individual logging into his or her workflow environment or the completion of several concurrent tasks.

Notification. An electronic message that is initiated by the occurrence of any other workflow event. Notification may act as a trigger but is usually thought of as an end step in the process.

Iteration or Negotiation. The repetitive execution of an object, using substantially the same rules each time. For example, a purchase order that must be signed by three managers is an iteration of the same object.

Duration. The time required to complete an event or the period of time for the deferment of a specific event until its dependencies have been satisfied. Duration does not include transfer time since in a workflow system information transfer time is eliminated and only task time remains. This means that events are always queued and waiting for information to process. It may, however, include the time specified within which the next event must be processed.

Dependence. This represents prerequisite objects that must initiate the dependent object. In some cases there may be numerous prerequisites to an object. For example, the approval of a purchase order by three managers before an order is initiated.

Completion. This is the last event in a workflow process for a particular object. The only exception is notification which may proceed as a result of completion. In a string of objects, completion of one object may lead to initiation of another.

Intelligent Documents: Another Approach to Workflow

The object-oriented approach used by many products can also be used by incorporating the object into the document. This approach does not provide all of the functionality of an overall environment for workflow that incorporates multiple office automation applications, but it does work well within a pure document management scenario. Intelligent documents use a combination of compound document structures and rules-based routing, processing, and modification. This may be ideal for a publishing or authoring application that involves workgroups. In these applications, the document must be transportable across multiple platforms in many cases. This is accomplished through adherence to document interchange standards such as ODA/ODIF, ODMA, Shamrock, OLE, OpenDoc, and document viewers based on forms of Postscript.

The concept of intelligent documents also applies to the broader process of linking applications through embedded tools for launching applications such as Microsoft Windows *Dynamic Data Exchange* (DDE) and *Object Linking and Embedding* (OLE). DDE, OLE, and Apple's Publish/Subscribe function provide environments by which to attach data and objects to the applications that created them. This is a useful function for integrating applications and information but, once again, it does not take the place of workflow, which also associates the other attributes mentioned with the information object.

Undoubtedly, workflow will slowly move into the realm of intelligent objects and documents. This is, however, a longer-term direction. In the short term workflow will evolve in several areas. Most noticeably lacking in current generation workflow products is the ability to simulate workflow procedures prior to implementation. Vendors will begin providing these tools as a necessary part of the business process redesign effort of which workflow is often part. There will also be development in the area of linking objects within workflow applications through system-level utilities such as OLE. This will be especially critical as the workflow is used to provide a metaphor for the varied office automation environments that many users are currently using. The most dramatic shift will occur in the next two years as workflow becomes the enabling technology for achieving the objective of electronic information processing. It will be the conduit that enables many predecessor technologies, those that promised paperless processing, to finally deliver on that promise.

Evaluating Workflow Solutions

Identifying the correct workflow solution for your organization may well be the single most important aspect of an effective workflow application. As with most new technologies the market for workflow is characterized by numerous product offerings and approaches. Any attempt to make sense of such apparent chaos is haphazard, at best, unless you begin to categorize the types of workflow products available and then prioritize their functionality according to your organization's needs and requirements. That alone may be a monumental task. This chapter is intended to assist the user in evaluating available workflow products by providing a list of key workflow components by which to gauge each product. Each item is described in order to familiarize you with the mechanics and the benefits of alternative approaches.

Keep in mind that not every item on the checklist will be found in each product, nor will every one be required by your organization. But each should be carefully studied and prioritized in order to select the approach that offers the greatest breadth of required functionality, flexibility, and potential for success.

Imaging Capability

Workflow is often associated with imaging. The two are, however, completely independent technologies. The dependency seems to exist due to the synergy between the two technologies when imaging is the driving application. In fact, it makes little sense to consider imaging without also considering workflow technology. The inverse is not always the case. There are many workflow applications that do not require imaging functionality. The documents that are to be managed by the workflow system may already exist in a word processor, spreadsheet, or database application. Before considering the merits of image-enabled workflow it is important to understand this distinction.

If imaging is an important part of the organization's existing information systems or a component which you are considering to enhance the organization's workflow, you should be careful to evaluate the workflow system from the following perspectives.

First, consider the extent to which additional software and hardware will be required in order to facilitate an image-enabled workflow application. For example, some workflow systems can easily handle the ability to address and view an image, but many rely on an underlying image processing application and hardware infrastructure.

Simply being able to retrieve the image will not ensure adequate response time, network performance, or memory requirements. Since workflow acts as a layer of technology above the underlying data management applications (DBMS, imaging, text retrieval, etc.), you must first establish the viability of these technologies and then move on to the implementation of a workflow application.

Second, you should evaluate the approach which the workflow product takes toward the integration of imaging. Often this will be a key consideration from the standpoint of ease of use. Users are typically in favor of an approach that minimizes the fragmentation of the interface environment. If they are already comfortable with Windows, an approach that utilizes a Windows-based imaging system may appear to be ideal. But that is not always the case. Even a Windows-based imaging system may require that the user learn three separate metaphors for the desktop—one for the Windows environment, one for the workflow application, and yet a third for the imaging system.

The best approach is to provide a single metaphor for the desktop that will be consistent across each of these environments. These products will provide a metaphor for the desktop that includes icons and document objects. Images and word processing documents can be invoked and launched from the desktop for modification or viewing, which is fine as long as the user's environment remains the workflow desktop. But once an imaging application is launched the user will have to be able to work with the imaging application that manages the images as well.

Third, don't let imaging become a prerequisite for workflow. Workflow can be designed, implemented, and used effectively without imaging. Over time you may decide to augment your workflow systems with imaging. It is always wise to consider the long-term flexibility of the workflow application for new data types, such as images, but you may simply not be able to justify workflow if it is burdened with the investment required for fully functional image processing.

Document Management

The distinction between workflow and document management software is often blurred. In many cases these two technologies can be used interchangeably to describe applications that manage the sharing of documents across multiple users and networked environments. The distinction is that workflow provides a means not only to track versions, ownership, history, and retrieval but also to productively manage the tasks that use the document. This enables the development of rules and sophisticated routing scenarios that are based on more than the document alone.

At the same time, many workflow products do not offer a fully functional set of document management facilities such as version control, content-based retrieval, security, and audit trails. Despite the apparent synergy that exists between workflow and document management, the two sets of functions are not always found in one product. If document management is a key concern for your organization you should consider the following.

First, determine the primary goal of the application. If it is the tracking of documents, as opposed to the proactive management of document-based tasks, a workflow application may not be required. On the other hand, if document management is a secondary concern, a few basic facilities may be sufficient. These would include the ability to assign ownership to a document, set security levels for access privileges, and maintain version control across a distributed environment.

Second, determine the need for a document management system. Document management may require the ability to assign fixed fields, such as author, date, version, and index number, to document information. This can be done through the added facility of a document management system, but in many cases it can also be accomplished through an integrated database and development language. Be careful not to underestimate the power of a fully functional document manager or to overestimate your ability to build this level of functionality in a cost-effective manner. In almost all instances you are much better off choosing a workflow package with basic document management functions rather than attempting to develop you own.

Office Automation Integration

Workflow applications are, by definition, integrated with other office automation tools such as word processing, E-mail, spreadsheets, and database. Since these tools are diverse and often work with separate collections of data, the workflow system must provide facilities by which to integrate with each OA application. This may be accomplished through an operating system function such as OLE/DDE, launching of the individual OA applications, or the user environment (i.e., Windows). The specific method best suited to your workflow application will depend on the sophistication of the users and the integrity required of the integrated data sets. The integrity issue is the most difficult to address.

Integrity refers to the ability of the workflow environment to ensure that data modification across several data sets is synchronized in such a way as to avoid the processing of incomplete or unallowable transactions. For example, a customer support representative cannot submit an error-resolution request to engineering until the customer support supervisor signs off on it and the customer's account is checked to verify that the maintenance agreement is current. This may involve three data sets: the error-resolution-request document, the fixed data required for the supervisor's sign-off, and the database used to store financial information about the customer's account. All three must be coordinated in order to process the transaction. The workflow application must, therefore, access, retrieve, and potentially update one or more data sets.

Business Process Automation

Since workflow is as much a discipline as it is a technology, you would expect the workflow vendor to also provide a methodology for business process redesign and workflow simulation. In all but a few cases you would be mistaken. Workflow methods are not prevalent at this time. Although a few vendors have developed their own proprietary methods for workflow analysis, most vendors will opt for a workflow analysis as part of the product offering. This obligates you to work with the vendor to define the problem and the potential solutions. If you are looking for a truly objective analysis of your problem and an objective assessment of the potential solution, you are far better off obtaining an education and an open methodology, such as The Delphi Workflow Design Method.

Off-the-Shelf Functionality Versus Customizability

Workflow products provide a range of capability from off-the-shelf user-oriented products to highly customized products that require a considerable amount of design and programming prior to installation.

The trend in the workflow market has been toward products that provide a combination of off-the-shelf and customizable functionality. As is so often the case, however, the best of both worlds is not always a union without compromise. Since most products have evolved from one of these two extremes, they will favor the off-the-shelf or the customized approach. The key to identifying the product best suited to your needs is establishing how much customization your users and applications will require within a given period of time.

Workflow products adopt one of three process models: *ad hoc, transaction-based,* or *knowledge-based.* Each of these has a distinct set of benefits. Transaction-based workflow products are designed to manage high-volume, complex tasks that are repetitive in nature. For that reason transaction-based workflow products usually provide high-level job definition tools and usually require a substantial amount of customization. The focus of these products is on performance and throughput rather than ease of use in job definition.

Products requiring a high level of customization are unsuitable for the ad hoc workflow environment since ad hoc workflow products are highly individualized in nature. The focus of evaluation for ad hoc workflow should be placed on ease of use and the effectiveness of job definition tools and the desktop environment. Products that emulate familiar desktop environments such as the use of folders and other desktop icons will be easier to use than those requiring a high level of training and customization. The ability of end users to create and modify workflow applications "on the fly" is essential in an ad hoc workflow environment. Job definition tools should be graphically oriented and require a minimum amount of training to use. Workflow job definitions should be easy to develop and modify, thus minimizing reliance on programmers and developers.

Some products will provide a well-defined interface and object definition tools that enable end users to create workflow applications without programming. These products enable the end user to define and modify the workflow process in this way. The problem with this approach may turn out to be that the object parameters and supported rules are not robust enough to handle all of the possible contingencies. Workflow applications that contain complex procedures and rules will require customization beyond that provided by most icon-driven front ends. Products that focus on complex workflow applications provide advance workflow definition tools and 4GL scripting languages to supplement basic off-the-shelf functionality. These provide limited user-definable workflow capability which can be enhanced through the use of advanced programming tools and APIs.

While many of the user-definable tools on the market are appealing, choosing one of these for a highly complex transaction-oriented application often sacrifices performance for the end user's ease of development. This could have a detrimental effect on the implementation and long-term success of the workflow application. Organizations considering a workflow application such as this should expect a high level of customization and should look for products which provide the highest performance available, while recognizing the need for ongoing technical support.

Scalability

The vast majority of workflow users will have to consider the migration of workflow across the enterprise. Few workflow applications will remain in a small workgroup for long. They quickly migrate into multidepartment and enterprise-wide applications supporting larger numbers of users. In this scenario, the off-the-shelf package that may be ideal for a small workgroup will soon be outgrown as demands of users begin to diversify and create the requirements for specialized components that are not supported in the existing product functionality. At the other extreme, it is unlikely that you can justify spending upfront time customizing a workflow application for small ad hoc applications involving a limited number of users.

Since few organizations initiate workflow on an enterprise-wide level, the ability to expand workflow applications is an important element to consider when evaluating a workflow product. You should be especially careful to select workflow products that provide open access to the documents and indices that have been used in the workflow application. This will enable you migrate while preserving the investment in your data and access paths. It is also important to consider the compatibility of the workflow system with standardized user environments, such as Windows, and adherence to image standards, such as TIFF and CCITT compression. If the workflow product is also integrated with a standard DBMS, you will be able to preserve much of the structured information relating to the documents and data used in the workflow.

Database Support

The database is one of the most essential aspects of any workflow system. It will provide the ability to integrate workflow with other applications tools through a common data-level layer. The database should support the ability to point to information in such a way that the workflow information can comprise a variety of data types. The database is also used to store workflow application procedures and rules as well as data on specific cases, providing an audit trail for all workflow activities and actions.

Most workflow products use standard third-party databases such as Oracle or Sybase. The advantage of this approach lies in the ease of integration and the ability to migrate the information to a different platform or system should the need arise. Products that contain proprietary databases generally provide good performance, but organizations that select this type of product may find themselves dependent upon the vendor for support and will have difficulty migrating data to a different platform should the need arise. If the data contained in the database must be maintained for a long term or if future migration or expansion is anticipated, selection of a product built on standard third-party databases is recommended.

Job Definition Tools

Job definition tools are the means by which a workflow application is written. The functionality of these tools varies significantly from product to product. Ad hoc workflow products usually contain graphical tools that allow end users to develop their own workflow applications even on the fly. These tools are useful for the development of simple ad hoc workflow applications but are limited in their ability to handle more complex applications requiring a high level of integration. At the other end of the spectrum are the highly transaction-oriented workflow products which contain a combination of graphical tools and 4GL scripting languages for use primarily by programmers for the development of customized applications. They are often difficult to use and require a high level of programming experience. They are also not easily modified and require ongoing programming support for the maintenance of the system.

Advanced Tools

Advanced tools are provided with many workflow products to enable customization of workflow applications. Tools range in functionality. Most products provide *application program interfaces* (APIs) but additional tools include: application program libraries, 4GL languages, and integration with standard application development tools such as Microsoft's Visual Basic and Sybase's Powerbuilder. These customized workflow applications provide a high level of integration with other business systems and legacy applications.

4 Imaging

The Origins of Imaging

Determining the reasons for imaging's fast rise to popularity does not require a great deal of scrutiny. The reasons are not surprising. Imaging represents a technology that uses a traditional page metaphor of the document. In most cases, imaging does not alter the size, shape, composition, or presentation of the paper page. In this regard, imaging is a direct replacement for paper-based document management systems. Therein lie its strength and its weakness. Imaging can quickly replicate a paper-based process in its flow and its use. But replication is often the most inefficient form of automation. Before exploring these issues in greater depth let's look first at the origins of imaging, its basic architecture, and its components.

Figure 4-1 shows that imaging had its genesis in widely different areas and technologies that have come together over the last decade to form what we now refer to as the imaging industry. Distinct industries such as aerospace, medicine, engineering, and manufacturing have made significant contributions to the component technologies of machine vision, analog-to-digital conversion, compression, transmission, and computer display devices. But these technologies were often relegated to isolated and obscure applications that hardly had an effect on the mainstream of office automation and the business users of computer-based information systems. The more significant events leading to imaging's thrust into the mainstream of document management were much more subtle at first, but ultimately the most pervasive influences on its rise to popularity.

The Imaging Metaphor

A key factor in imaging's rise in popularity during the last decade is the revolution in telecommunications and the resulting transformation

1925, AT&T service to send wirephotos

Micrographics

Pioneered by Kodak in 1930s

Satellite technology

1960, NASA contracted JPL to develop system to convert satellite video to digital form telecommunications

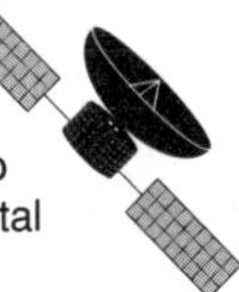

Audio/entertainment technology

Videodisk developed by Sony in early 1970s
CD developed by Sony and Phillips in late 1970s
4.75" platter with 650 MB

Figure 4-1 The origins of imaging. Although viewed as a relatively new technology, imaging can trace its roots back as far as the 1920s and to many related areas of technology. Advancements made by AT&T in the form of transmitting wirephotos were an early form of facsimile technology, a form of document imaging. Likewise, Jet Propulsion Laboratory's (JPL) work with space photography was a precursor to today's document imaging systems. Though it was not a digitized form of data, general user acceptance of microform ushered in the concept of storing documents in a media other than paper, and has helped acclimate many organizations to the concept of document imaging. Similarly, the viability of document imaging was strengthened by the advent of optical storage, which had its genesis in the entertainment industry.

of business communications through a commonplace device simple enough to use for even the most computer-illiterate office workers—the *facsimile* machine. Facsimile transmission is an old technology that only recently rose to stardom. In fact, the first facsimile machine was in use as far back as 1925! It was a crude version of its modern-day counterpart, but its developer, AT&T, could transmit photographs over standard telecommunications lines. Today a business does not exist, no matter how small, without fax capability. Many people think nothing of using fax at home, in their automobiles, or even in commercial airliners using their portable computers or hand-held personal assistants.

The rapid and sudden proliferation of facsimile is a phenomenon that we often take for granted, but it begs us to ask why it took so long to commercialize something which, in concept, had been available for several decades. We could claim that it was the dramatic decrease in the cost of technology, or the ever-increasing burden on business to communicate in a global market, or perhaps the increasing availability of desktop computer systems. Then again, who's to say it wasn't the raised expectations for immediate document delivery through the proliferation of overnight delivery services? But all of these reasons would be irrelevant if not for the development of one fundamental item—a *standard.* It was the universal agreement on a standard for communication facsimile that suddenly gave us the global medium by which we transfer documents electronically.

The CCITT standards for fax (GIII and GIV, for Group 3 and Group 4) provided the security for both developers and users to invest in the technology. Whether any standard itself is good, mediocre, or poor from the standpoint of technology makes little difference in the world of mass markets. The simple fact is that standards, formal or de facto, propel a market further than virtually any technology on its own merits.

In the imaging industry these standards are easy to come by, which is precisely the problem. In the area of optical disc technology there are currently more than 20 standards for six completely different media. Even image files that use the popular *tagged image file format* (TIFF) can actually be in any one of 16 common TIFF formats. This state of flux has hindered the adoption of imaging across many enterprises using multiple imaging solutions.

Sony's Black Eye

A dramatic example of the power of standards can be found in another industry that has contributed a great deal to the emergence of electronic imaging—the entertainment industry. Audio entertainment was the primary thrust behind the development of optical storage, from which CD-ROM evolved. In fact the giants in the entertainment industry, companies like Sony, learned that lesson well, as did many consumers who purchased a Beta VCR.

Sony's defeat in the VCR standards battle surprised many technologists who believed, correctly, that Beta was the better technology. But it did not survive in the mass market. Sony did, however, capture the professional video market and Beta has become the de facto standard of all studio videography. That may not appear to be a dramatic revelation since, after all, many products and technologies can claim to be better yet not be a leader in their field. But Sony, the company behind Beta, was a major player with significant resources and clout in the consumer market. That episode marked a turning point for Sony, and in many ways for the information industry.

Beta was a power technology for power users. Unfortunately only a small segment of the market for VCR technology was comprised of power users, as evidenced by the fact that less than half of all VCR users know how to program their VCR. The majority of VCR buyers were end users who needed the convenience and security of a compatible solution that would work with any video purchased at their local corner video outlet. End users choose a technology, most often, based on their perception that a large number of other users have also decided to make the same investment. That principle applies to any technology that hopes to make it into the mainstream of the broad-based, end-user market.

Today's battles for standardization are being fought on the desktop, at the point of the end user. That is especially true of a technology such as imaging, which replaces one of the most personal forms of information—paper.

Micrographics

Another area of technology that has had a significant impact on the evolution of imaging is that of micrographics. We will not discuss the topic of micrographics in this text.[1] It is, however, important to acknowledge the past and present role of this technology in document management systems. Micrographics represents a significant legacy of images for many organizations, and especially so in industries such as banking and libraries. Many of the battles being fought today over the legality of images and their admissibility as evidence also had to be overcome in the micrographics industry, as did the problem of changing metaphors for workers, and the justification of document conversion. The lessons learned are valuable to consider in the face of many concerns that organizations have when migrating to imaging. The most significant change is the rate of change itself. Electronic imaging is providing a means of converting and utilizing images at a pace that

[1]Many outstanding texts have been written in the field of micrographics. This text is concerned with electronic documents and although micrographics will undoubtedly continue to be a document management means for some time, especially in the archival area, it does not represent an electronic document medium.

eclipses that of the micrographics industry. Establishing standards for the use and legality of electronic images must happen over an exceptionally compressed time frame when compared with the decades of micrographics evolution.

Imaging and Document Management

A great injustice has been done to evaluators and users of document technologies such as imaging. They have been coerced by throngs of well-intentioned imaging vendors into believing that documents consist of only one data type, either image or text, when in fact, even our most casual interactions with documents include a variety of data types. Consider the universal metaphor of the newspaper. Graphics, illustrations, charts, and photographs are woven through text in a rich tapestry of meaning. The relationships expressed by the arrangement and interaction of various data types have value to the end user. In a business process, documents are arranged, organized, and processed as compound objects—you need only look as far as your desktop, and the endless stacks, paper clips, staples, and dog-eared papers for testimony.

Documents are compound by their very nature, as are the business processes they support. Limiting a document's electronic form to only one information type (whatever type that may be) not only creates an artificial environment, but also limits the utility of electronic document management systems by forcing users to interact with multiple applications, operating environments, and tool sets to access all of the document information needed to perform a task. Consider the number of document windows you may have open at any one time for a simple task such as customer support. Customer information may come from a database record, correspondence in a word processing application, an image of a customer's handwritten note in yet another application, a spreadsheet with account balance information, and an E-mail connection with customer communications.

A true compound document architecture would provide the foundation for creating applications that act as a central point of integration for processes involving multiple information sources. It is likely that at some point in the future, the concept of applications and operating systems and environments (an operating environment is best likened to products such as Lotus Notes which sit on top of operating systems) will be completely obscured from users. You will no longer launch a word processing application, or a spreadsheet, or a database. Nor will you traverse endless arrays of windows to coordinate all of the information you need to support a task. The compound document will become the metaphor for work. Information will be embedded into documents in such a way that the associations of one piece of information to another are clearly evident. These documents will represent the embodiment of process knowledge as well as information objects. They will replace the desktop, and remove users from the enormous administrative burden of file management—defining workflows, and application dependencies. If that seems far-fetched, take heart. It is. No one said it will happen tomorrow, only that it will inevitably happen. In the interim, we will have to deal with a variety of obstacles that stand in the way of this utopian view: obstacles which are firmly rooted in the convoluted reality of today's myriad platforms, applications, and standards (or lack thereof).

The Compound Document Fiefdoms

Capturing and maintaining the fundamental value of documents, their compound nature, has been an elusive goal for electronic systems until recently. Newly proposed standards and architectures are emerging that promise to make the compound document a reality. The problem is that there is no single architecture, as of yet, that will be a clear leader. A variety of organizations, coalitions, and existing standards are vying for control and in some cases collaboration over compound documents. The most visible contenders to date are Microsoft's OLE, the Object Management Group's CORBA, Apple's Publish and Subscribe, OpenDoc, Shamrock, ODA, and SGML.

Each of these attempts to address two basic aspects of a compound document: structure and organization. (See Fig. 4-2.) Structure refers to the way in which the components of a compound document relate to one another within the document. Structural components may be the title, summary paragraph, footnote, author's name, and date. Most structure is represented by tagging the document with special markup codes that can be interpreted when constructing or retrieving the document. Organizations refer to the relationship of a document to other documents and processes: a document may be part of a given process at a specific time, or it may be related to another document based upon a certain condition. For example, if the call is from an existing customer, you need only to look up prior customer correspondence.

When the structural elements of a document are combined with the information it contains and then linked with other compound documents, a value chain is created in the organization. This value chain reflects the business processes of the organization as precisely as an

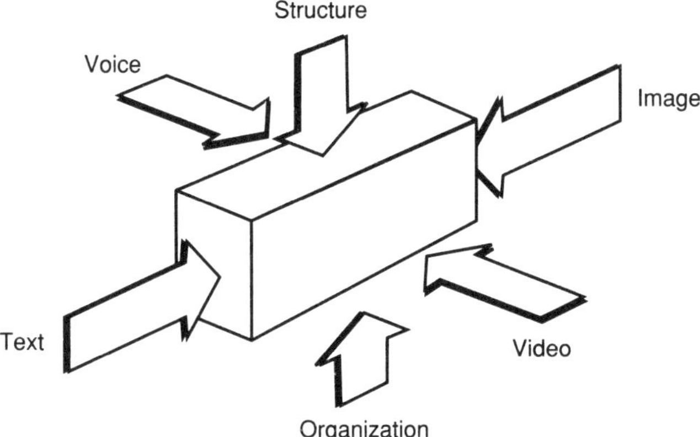

Figure 4-2 The components of a compound document.

assembly line reflects the manufacturing processes of a factory. The obvious difference, however, is that the business process must respond to change much faster than a factory assembly line. A compound document management system that integrates information as well as processes intelligence can help support an agile environment, which can be quickly reconfigured to adapt to changing competitive, economic, and market forces. (See Fig. 4-3.)

The Standards

The variety of standards, existing and proposed, that are attempting to define compound document interoperability is perhaps the biggest contributor to the evolution toward compound documents; it is also the

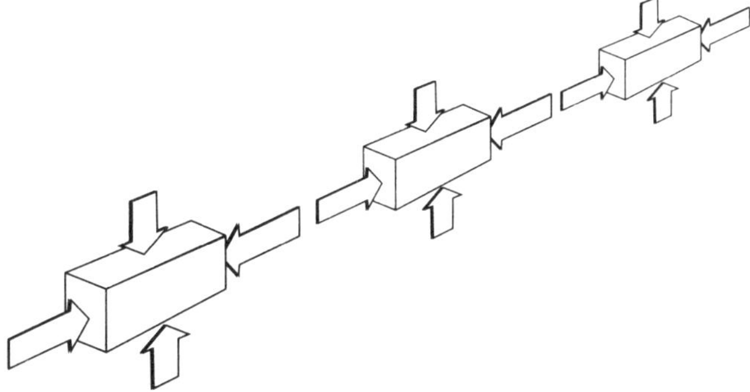

Figure 4-3 The adaptive organization is supported by compound documents. In an adaptive organization compound documents help to quickly retool business processes.

greatest liability. Since compound documents reflect a business process, they must be able to follow that process through and even outside of the enterprise to suppliers and business partners. The most popular of these standards are SGML, ODA, OLE and CORBA, and OpenDoc.

SGML. *Standard Generalized Markup Language* (SGML) has become the hottest topic in the document management arena today. Markup languages have been around for some time in the publishing community where they are used to define the various components and characteristics of a structured document. SGML offers the ability to identify document sections in a such a way that a document can be retrieved based on these structured items across a variety of SGML-compliant document platforms. Although the markup process may at times be a tedious one, there are automated tools and libraries of document type definitions (DTDs) available for certain standard documents such as those used in the Department of Defense Computer Aided Acquisition and Life-Cycle Support (CALS) initiative, and the benefits are enormous for large structured document collections. SGML could be extended to include any number of data types including images, voice, and video, although today it is used primarily for text.

One of the problems with SGML, however, is that it does not provide a facility for associating foreign applications with the information objects it defines. In other words, the SGML document has no application intelligence. It is simply a means of identifying the structural components of a document. (Newer forms of SGML such as HTML (Hypertext Markup Language), used widely on the Internet, do support application links.)

One interesting aspect of SGML is the ability to create *virtual documents,* which are spontaneously created subsets of much larger document collections. The virtual documents are created when a user requests information through some type of query. An example may be a query of the IRS tax code for "allowable deductions for dependents." At least 50 separate sections of the tax code refer to issues related to deductions for dependents. By linking the SGML tags to a database that identifies the structure of the document's subject matter, a subset could easily be extracted and presented to the user as a mini tax code prepared specifically for this query.

ODA. *Open document architecture,* also called *office document architecture* (ODA), is similar to SGML in that it provides for structural identification of document components. ODA is often associated with *Open Document Interchange Format* (ODIF), which is a specific implementation of the ODA markup. Although ODA has been around since the mid-1980s, it is not yet widely used by document vendors. In fact, ODA's evolution turned out to be indicative of the single greatest problem with establishing common document architecture for compound document transportability: vendor cooperation. ODA promised to create a standard metaformat for all document types. It included text, images, spreadsheets, database tables, voice, video, and according to one early specification, even holographic images. The best way to envision ODA was as a railroad roundhouse. Documents would go in from any vendor's platform and end up on any other ODA-compatible platform. Vendors eagerly signed on and soon a small cadre claimed support for ODA. Unfortunately, railroad roundhouses work only if tracks lead in as well as out. In the case of ODA, a significant number of vendors were prepared to accept documents from the format standard but were not prepared to export to it.

In 1993 a group of vendors gave ODA new life by forming the *Open Document Architecture Consortium* (ODAC). This group created a resurgence in ODA within Europe, but has not yet had a widespread impact on the U.S. market.

OLE. *Object linking and embedding* (OLE)) is not often thought of as a compound document architecture since it addresses only one platform. Although it is limited to Windows applications at the present time, Microsoft's plans for client/server computing may turn OLE into a universal-client metaphor for compound documents. Among other things, ODA provides the ability to place information objects created in one OLE application into any other OLE application. For example, a pie chart from a spreadsheet application is placed, the information object is moved, and the associated application is linked to the object. In this way, an end user can invoke the information object directly from any OLE-compatible application and launch directly into the spreadsheet which created it.

Although OLE is a substantial step forward, it does not solve the problem of object containers (the user repository of document objects) since it requires that a specific application act as the front end for the compound document. You have to choose a word processing or other application suited to your presentation requirements as the repository of your document objects. This defeats the purpose of application independence.

CORBA. A competing standard, at least at the time of this writing, is *Common Object Request Broker Architecture*. Backed by Apple, Borland, and WordPerfect, among others, CORBA provides a high-level container within which any variety of a compliant application's information types can be embedded. This offers a truly application-independent approach, at least to the extent of those applications supporting CORBA. One way to look at this is to view the document container as the familiar desktop, where multiple windows and multiple data types occupy a common space. Instead, CORBA may lead to another common space—the document object.

OpenDoc. Apple's and IBM's response to OLE is *OpenDoc.* OpenDoc uses the CORBA approach of putting applications and information into a container. Unlike OLE this container is an application-independent, sophisticated, compound document architecture which includes documents, parts, and frames. This approach places a strong emphasis on object orientation. Parts are individual information and application objects within a document, and can themselves contain other parts. The parts are separated within the document by "frames" which are logical structures representing the processing and organization of the document. In Apple's view, standards such as SGML could be implemented as an application-specific function and have no effect on the compound document architecture. This allows for the existence of multiple standards in a single architecture.

Levels of Object Intelligence in Compound Documents

Markup (no intelligence)—cut and paste, most windowing environments, SGML.

Live Links (application intelligence)—OLE, Publish and Subscribe CORBA.

Object Integration—none today; look to workflow and operating environment vendors (i.e., Lotus, Novell, Microsoft, Apple) to provide these.

Shamrock. One of the newest standards proposed for document interoperability is that of the Shamrock Group. Shamrock was formed in late 1993 and announced it would establish a set of APIs that would allow multiple document applications to share objects in a common repository structure. Admittedly, the group is barely out of its embryonic phase at the time of this writing, but it does provide added emphasis to the importance many vendors are placing on the ability to transport complex documents through the enterprise. It also attests to the dissatisfaction of vendors with the array of standards already available.

Open Document Management API (ODMA). ODMA announced on July 11, 1994, the release of a set of standard APIs for document management integration with other applications, including workflow. The purpose of this standard is to allow desktop applications to work with ODMA document management systems. The group, which includes such vendors as Novell, Soft Solutions, PC DOCS, and Documentum, is actively moving ahead with the introduction of standards. Their fast pace may be a defining factor in their success.

Ultimately, Compound Documents Are the Desktop. Where are all of these standards going, and how will you make a decision about a compound document architecture? For the time being, it is best to rely on standards that preserve the links to a document object's native applications. Facilities such as OLE provide this, although they lack the process intelligence needed to provide integrity for a document's use. The best way to look at the benefits and liabilities of the approaches presented, and the many more offered by individual vendors, is to keep in mind two basic tenets of document management: information assets and process assets. You must preserve both across time in order to migrate your documents from standard to standard, and application to application.

Preserving your information assets means using approaches that do not disturb the ability to access document objects in their native form. For example, images remain in CCITT G4 format, text remains in ASCII or some format transportable to ASCII, indices to documents and document objects remain in standard DBMS formats (SQL) or flat tables, and applications remain attached to their associated document objects. Process assets represent the rules, roles, and routes that define the value chain process in your organization. Preserving these means once again maintaining an open versus proprietary repository for document communications, workflow, and workgroup applications.

The irony is that the compound document standards which we are today trying to fit into conventional application and operating system desktop paradigms may end up becoming the desktop paradigm. When all of this has been done, enormous change will have occurred that will dwarf our current expectations for compound documents. Your objective is not to stem the tide of new technologies and standards, or even to anticipate them, but to preserve the fundamental components of your documents and your process across change.

The Imaging Market

Imaging encompasses applications that span virtually every industry and discipline, not just business documents. Engineering documents, CAD, scientific and medical documents, x rays, and commercially published documents could all be considered imaging applications. Somewhere we need to draw a line if we are to have a meaningful discussion about imaging. That line cannot be drawn by industry since each of the applications we have just described could be used within a number of industries. Instead we will draw the line based on the way in which the imaged information is used.

In applications such as medicine, space exploration, and in some cases engineering, the priority is on enhancing the image, not on adding value to the process of which the image is a part. In our discussion, we will focus on a different usage or direction for imaging technology: the priority of adding value to the business process by virtue of the image.

This usage of imaging shifts the focus from the image itself to the means by which we attempt to add value to the business process. This is a key distinction because the marketplace for that niche is the most radical departure from the paper-based image. Images that are used to facilitate business processes traverse many organizational boundaries. The faster they execute this traversal and the greater their integrity across these boundaries the higher the payback to the business process. Ultimately we could apply that principle to almost any use of images. An x ray, for example, is much more effective if it can be accessed by a requesting physician on demand. If the x ray is stored in an off-site archive, however, retrieval and routing to the physician could take hours or days. An electronic image could be retrieved in seconds or minutes. This causes a rethinking of the priority placed on the use of the image itself. Consider, for instance, whether it is more effective to have an x ray that has marginal quality delivered to the operating room in seconds or to wait hours for the best possible resolution, which happens to be on the film stored in a warehouse. These are not moot questions. Indeed, imaging is causing many users of imaging to reconsider the fundamental tenets of their process and priorities.

With this as our definition of the imaging market, we state that the size of the imaging market is hovering at about $3 billion.[2] These numbers will certainly vary, depending on whom you ask. But what is striking is the inability of the imaging market to keep pace with the fantastic predictions that have been forecasted over the past several years. In fact, if you compare industry forecasts for the last four years an interesting phenomenon occurs. Each year the forecast for the size of the imaging market has gone down by approximately 30 percent. With discrepancies of this type, it is amazing to consider the enormous investment made by existing vendors and new entrants.

The imaging market has been shaped by what economists refer to as a supply side model. Vendors have spent untold millions over the past

[2]Size of the imaging market in 1994 (Delphi).

several years trying to prime the pump of industry and government by creating an awareness of imaging's benefits and payback.

When you have a market expanding this quickly, chances are that mind share grows much faster than market share. And mind share can be very deceptive. Although interest in imaging has certainly kept pace with even the most optimistic expectations, organizations have not been as open with their pocketbooks as they have been with their minds. Before running for the door, most organizations are stopping to see if they can smell the smoke in the air.

Over time the imaging industry has become more realistic about future projections. The reason for all the hyperbole is simple. Imaging has been a technology of great promise. The substantial investments in time and money made by imaging vendors and associations to promote the benefits of imaging and to bring attention to this new technology have worked. Claims of stellar growth rates and commensurate payback have caused many to look carefully at the benefits of imaging. Now, however, the scrutiny on vendors and products to perform up to expectation is mounting. The marketplace, investors, and business planners are looking carefully at the industry's ability to meet projected growth rates. Now that attention is being paid to imaging it is obviously preferable to make conservative projections that may be easily met and exceeded rather than to fall short of inflated projections.

Additionally, the general move toward desktop commodity imaging solutions has made it easier to buy into inexpensive solutions that test the benefits of imaging before committing to it. This is causing the imaging industry to switch to what economists refer to as a demand model. But in this new marketplace, based on low-cost entry-level solutions, many are asking what role imaging's pioneering organizations, which had focused on large-scale enterprise imaging applications, are going to play in the future of imaging.

Indications are that these vendors are more than capable of providing competitively positioned solutions, if needed. In fact, many of these vendors are willing to offer third-party imaging systems through other imaging vendors. Their emphasis is shifting from pushing a single all-purpose hardware or software solution to that of adding value through the process of integration and services.

Although commoditization of hardware and platforms is inevitable, the additional cost of software customization should not be trivialized. The real question is not how much will it cost to customize, but rather do you want to buy experience from a third party and put the cost on your company's books, or are you going to absorb the cost with your resources. Either way there will be an additional cost. In fact, as discussed later in this chapter, implementors of imaging systems report that customization and integration costs account for anywhere from 30 to 60 percent of total system costs.

With this in mind the major imaging vendors are emphasizing a business model and service offerings that address the need for planning and integration. The impact this will have can't be underestimated. Traditional hardware vendors such as Digital Equipment Corporation can already be found recommending other hardware platforms and solutions to customers. It is not a novel concept. In 1992 Italian computer giant Olivetti announced that over 60 percent of its revenues were generated through service-related activities.

The percentage of service revenues and margins will increase across the industry as hardware and software margins fall. That should be no surprise given the decision of many hardware vendors to discontinue manufacturing operations and focus strictly on services and software.

But this also means that the low end of imaging systems for small workgroups and commodity imaging will belong to some other player that is yet to be decided. It is unlikely to be an organization that rests on the value added by services and integration. It's early to predict precisely the impact this approach will have on the market for imaging solutions, but it will undoubtedly push further toward open systems and document standards that encourage interoperability of imaging systems. This is a requisite for the emergence of a true commodity solution at the low end.

The process of selecting an integrator will certainly become much more interesting since virtually all major hardware vendors will be competing for the service-value-added component of large-scale imaging opportunities. If the thought of a vendor such as IBM becoming a service provider is difficult to envision, it should not be.

All traditional hardware systems providers will have to adopt this same model in order to deliver value for their solutions. Although these large organizations do not turn on a dime, the unavoidable reality is that hardware can no longer be the differentiator. The demands of large-scale imaging systems with thousands of users may be met best by the mix of platforms and services that a large vendor can offer. But the hard components of imaging will quickly turn into low-margin commodities, and desktop hardware will be more than capable of handling everything from single-user to enterprise imaging solutions. What will remain is the value added through expertise in the process. That is what these vendors have acquired during their many installations of imaging. The translation of that experience into software solutions is the principal value they bring to their customers. That experience, whether from vendor, consultant, integrator, or first-hand knowledge, will continue to be the most valuable element of cost-effective imaging systems.

Imaging for the Masses

Imaging is no longer a showcase application of the *Fortune* 100 and those organizations brave enough to undertake massive reengineering efforts. Plummeting software and hardware costs are on the brink of creating a commodity market for imaging and although still a consideration, the cost of imaging is no longer the key obstacle for most evaluators. According to a Delphi survey of 200 evaluators of imaging, organizational and technology issues were considered to be greater impediments to imaging than was cost by a 2-to-1 margin.

The latest breed of low-cost imaging products speaks to the value of an incremental approach to implementing imaging systems, something that was once thought of as anathema by many imaging evangelists. These evangelists believed that the more you spend and the more you reengineer, the higher your payback. There may have been shreds of truth to that message for organizations crumbling under the inefficiencies of their existing business process. But the majority of potential imaging users could not afford to turn their business model inside out for an unknown payback. These organizations repeatedly cried out for solutions that could be implemented with minimal investment and risk and yield a reasonable payback.

At first, the call for lower costs fell on vendors with deaf ears. Multimillion-dollar imaging integration efforts that came to be known as *enterprise imaging* were sufficient in numbers to fuel the infant industry. In that frenzy of opportunity and interest, a market with enormous potential was shaped. But, year after year, projections for extravagant growth rates fell short of reality. In fact actual enterprise imaging installations among even the most successful and better-known vendors numbered only in the hundreds. It soon became clear that the big boom had the potential of turning into a bust. The recession was blamed, of course, as was the lack of understanding of the potential benefits. After all, who wouldn't invest millions in new technology if they really understood the payback?

Enterprise imaging, at that time, required dedicated hardware, extensive customization, and a high level of financial and organizational commitment. That was much too expensive and drastic a proposition for more than a handful of organizations. But organizations such as American Express and USAA established the viability of large-production imaging installations and more importantly established the basic criteria for payback. One of the most important of these was that of reengineering the business process to take advantage of the new technology and the opportunities it presented. As the visibility of imaging grew, it was the enterprise installations of this sort that gave rise to the popular conception that imaging had to be justified at a high level with substantial commitment of executive management in order to succeed.

During that same time, however, a small cadre of what where thought to be insignificant vendors, by comparison with the ranks of the major players, were busy developing and installing relatively low-cost imaging systems intended for much smaller implementations in workgroups and local area network environments. As the limelight was focused on reengineering for enterprise applications, start-up companies were busy addressing the ripples of new and slowly emerging opportunity left in the wake of increasing unrest among the ranks of end users.

These companies anticipated the appeal of imaging to small workgroups and individuals along with increasing power and connectability of networked PCs as a viable imaging infrastructure. Their push was to make imaging a commodity application across all functions of an organization. The target applications for this new genre of imaging were far less strategic then the mission-critical enterprise applications. They were typically workgroups with the basic need to share unstructured document information among users on a LAN. In sharp contrast to the showcase enterprise imaging applications, these were operational applications of imaging with low glitz but high value. The hope of this new breed of low-end vendors was that imaging would become as ubiquitous as the PC itself. That bet is now starting to pay off.

The Cost of Imaging

When Delphi surveyed organizations purchasing imaging systems in 1991, we found that the average cost of an imaging purchase was $600,000 for a *Fortune* 1000 enterprise. The cost of a pilot application averaged $200,000. In 1993, only two years later, the average cost of an imaging purchase dropped to $200,000 and the average price for a pilot was less than $50,000.[3] Today imaging is available at less than $100 for desktop functionality that rivals even the most robust enterprise imaging. By the time this text is read it may well be that imaging

[3]Delphi 1993 imaging market report.

is being given away with a variety of networks, operating systems, and messaging applications.

The reason for this plummeting of prices appears to be evident: downsizing of imaging to work on more powerful microcomputers. But closer inspection shows that although the price of hardware and software infrastructure is going down, the cost of integration and customization is remaining fairly static. Anywhere from 30 to 60 percent of a typical enterprise imaging solution's cost will go to integration services. The question, of course, is what this 30 to 60 percent of the system cost buys. You're paying to ensure the interoperability of the many hardware and software components involved in an imaging system implementation, and for an analysis of the business process that will be affected by the imaging system. These are two distinctly different aspects of an imaging application, so much so that it is often best to have each performed by a different service provider.

Integration can include any one of the following: simple setup and installation of the imaging system's components, customization of interfaces, development of a workflow for the electronic documents, upgrading of existing hardware and network infrastructure, education, document conversion, indexing of documents, and system administration.

What many people often ignore is the physical reality of infrastructure upgrades, which alone may represent one-half of the integration cost. These may include large-screen monitors, compression and decompression hardware and software for image display, higher-bandwidth networks, increased storage capacity, additional dedicated servers, and larger-capacity backup facilities. Additionally, basic requirements for expanded desktop memory and processing speed may render current desktop platforms inadequate for production imaging. All of this can add up to a fairly high additional cost per seat for the imaging system.

Conversion Costs

Yet, these figures still do not take into consideration the most important component cost to consider—document conversion. In many cases document conversion can more than double the cost of the imaging system. On average it is safe to assume that it will cost about $1 per page to go through a legacy document conversion process. That may seem high when you consider the high throughput of many scanners and the dramatic reduction in scanning hardware costs. But before you accept that premise, stop and walk around the file room or the warehouse where your legacy documents are stored. Look at them carefully—very carefully. How many page sizes are there? How many different color pages are there? How many documents are third- or fourth-generation copies? How many documents are stapled or paper-clipped together? How many documents have faded? How many documents have been crumpled or otherwise damaged in storage? The realization slowly sets in that there is no such thing as a homogeneous document collection. Such is the nature of virtually all paper-based document collections. As a result there is typically a fair amount of manual effort involved in converting any paper-based repository to an on-line repository, which includes but is not limited to categorizing and preparing the documents and proofing and indexing the electronic documents. Add OCR to the process and the cost can jump fivefold!

Be leery of studies that show minimal document conversion costs of $0.10 per page or less. These studies are based on document collections that have already been categorized, can be automatically indexed, involve little or no OCR processing, and consist of 8½-by-11-inch sheets of "clean" paper. That is rarely the case across existing document repositories until a thorough audit has been undertaken and a relatively small portion of the legacy is identified as an economically viable conversion base.

When to Invest

Conversion costs aside, the fact remains that the overall cost of an imaging system is declining. Given this steady decline in price, you may think you should wait until the price goes down. From a competitive standpoint, you often can't afford to wait. If your organization decides to postpone an investment in imaging, awaiting a lower cost of entry, the organization may simply not be here in three to five years. But it's not an either/or scenario. Nothing that we talk about in this book makes the case for an immediate all-or-nothing conversion of your documents to an electronic system, image or otherwise. There is a broad spectrum of alternatives that should be considered and understood before making that decision. Even then there will be a significant learning curve that organization will have to undertake in order to determine the best application of EDMS technologies.

Secondly, the market is slowly coming around to a clearer definition of standards for imaging. These will evolve as the industry matures, causing costs to drop even further, but more importantly will define secure investments for large enterprise systems. Starting a pilot implementation today offers your organization a significant advantage in balancing the benefits of an education and point solution of imaging with the ability to keep a keen eye out for developments in the industry as it evolves.

Integration is becoming less of an obstacle as vertical applications developed for specific industries, such as insurance, pharmaceutical, and manufacturing, are helping to alleviate the burden of ground-zero integration. Standards efforts of organizations such as TWAIN, previously called the *Connecting Link Application Source Peripherals* (CLASP), are making the integration of scanning and recognition peripherals simpler. Compression and image storage formats are securing investment in electronic image repositories.

As efforts such as these progress so too should your organization's exposure to imaging. You don't take organizations and turn them up on their earlobe. It just doesn't happen. You have to work incrementally toward an enterprise solution. Consider the following scenario: you begin the process by evaluating the technology, which takes, based on the studies we have conducted, six to eight months. Then you implement the pilot application, which again will take from eight to twelve months. You've already invested two years at that point and you've got about a year ahead of you to flesh out the application and learn from your users what's a reasonable rate of change and what's not. Three years are behind you at that point. Standards have not stood still and neither has the technology. With an obsolescence cycle close to 12 months, a decision to proceed with an enterprise imaging system from the outset without prior organizational experience could be disastrous. The result is an obsolete system before it has been fully implemented.

Today you are ideally positioned to take advantage of an incremental approach to implementing imaging and in two to three years you will be ideally positioned to undertake enterprise imaging. The key to making this work is, of course, selecting the most appropriate pilot application for the imaging system. Although we will look at this much closer in the chapter on document systems design, we will first outline the specific issues involved in justifying an imaging system and selecting an imaging solution and pilot.

Making Imaging Cost-Effective

With all of the activity in new product developments we have to wonder if the commoditization of imaging in the form of lower costs and easy integration will make solving the fundamental problem of document management, which imaging enables, any easier. And if it does, is the initial low-cost workgroup solution scalable into a cost-effective imaging system for the enterprise? Vendors of low-end imaging products certainly want you to believe that it is, and in many cases they may be right. The answer, however, is not that simple.

The key to cost-effective imaging is choosing a solution that has a measurable payback and the ability to grow with your organization's demands (or if it can't meet these demands to provide a path for migration). That will require an enterprise vision and an overall context for reengineering existing document management. This presents the most striking contrast for buyers of imaging systems: the availability of a low-cost entry-level system does not alleviate the need for thorough problem definition and subsequent administration of the system. In fact, because of their diversity, imaging solutions can lead to even greater fragmentation of information systems and therefore demand an element of centralized coordination.

The decision to invest in a particular type of imaging product should follow a careful evaluation of the full spectrum of costs involved. Surprisingly, in most imaging applications these may not be radically different for the software and hardware components of several alternative solutions. Many traditional high-end vendors are moving to a user-based pricing model that helps to equalize the cost of the software for medium to large workgroups. The more significant cost, however, will result from the analysis that will need to take place and the level of integration required.

Although every imaging vendor may be providing solutions for a commodity hardware platform, some products are clearly less customizable than others. Deciding which category of imaging product your organization should invest in requires a thorough understanding of your budget constraints and payback requirements. If both are moderate, there are instances where simply eliminating the inefficiency of paper can be a significant improvement in productivity. Entry-level tools can help in that regard. They may also provide an introduction to the benefits and metaphors inherent in imaging. Watermark, for example, provides easy integration with Windows-based applications through its seamless support of DDE and OLE. Packages such as OCR, text retrieval, and basic workflow provide a complete environment for imaging for a minimal investment. If you take this approach be sure to use a technology that is open enough to allow for the migration of images, retrieval indices, and process models to a new system. Locking yourself into a proprietary model may require reinvesting time and effort to reconstruct the imaging environment. Imagine for instance, having to reenter the index keys for 10,000 images because the database used to store them in one product is incompatible with another.

Also, consider the benefits of expanding the application to include basic workflow, OCR, fax, and content-based document retrieval. That will probably add little in the way of cost but significantly increase payback. Most packages will provide these either as add-on extensions or as standard functions.

An acceptable imaging system must also be responsive to the user's operational requirements for speed and ergonomics. In some cases, this may have less to do with product cost, networks, and user interface components than it does with the underlying technologies of compression and decompression, caching from optical disc, and image display devices.

One of the most important aspects of cost-effective imaging, and one that is often ignored, is the attention to basic display requirements. Investing in a full-page or dual-page display with a minimum of 120 DPI resolution can be crucial to user acceptance. Small screens, anything less than a portrait monitor, simply don't provide the real estate needed to manipulate multiple documents. Twelve-inch screens are fine for data entry forms and structured information, but they are a poor excuse for the user's old document management metaphor, the 3-by-6-foot desktop. I often tell people that the success of their imaging implementation can ultimately be measured as an inverse ratio to laser printer output. There is nothing more discouraging than investing in an imaging system just to find users stashing even more paper into their already burgeoning filing cabinets.

Selecting Your Pilot Imaging Application

Envision the implosion of a building. If you have ever witnessed this event you know that it is an amazing sight, not easily forgotten. Implosion is one of the most dramatic methods of rebuilding in a congested urban setting. Old high-rise structures are razed by using strategically placed charges of explosive throughout the building's structure in such a way that their detonation brings the building down upon itself. The result is a pile of rubble that takes up little more space than the original footprint of the building. Most amazing is that onlookers and structures just a few blocks away are not affected by anything more than a light covering of dust.

The collapse of a business cycle within an organizational setting is similar to this implosion of a building because you are not just streamlining the existing structure (the business cycle), but are razing the structure and reconstructing it. If you attempt to implode an entire organization in the same way, however, you are going to have to contend with a lot of upset users, a lot of frustration, and a lot of heartache. You have a remote chance of success only when you have a sponsor who is positioned in the upper echelon of the organization—a CEO, a chairperson—someone who has made it his or her life's ambition to change the way the organization is run. This is a pretty rare occurrence.

So how do you implode existing structures and replace them with efficient structures without having that level of sponsorship? One way is to segregate the organization into smaller business cycles. There are business cycles in accounts receivable, in accounts payable, in claims processing, in customer support. By looking at each business process as a series of incremental efforts at rebuilding the organization it is possible over time to undertake a massive restructuring without the risk or commitment required for a day-one demolition of the entire enterprise. This is almost always the case, even in the showcase examples of imaging implementations.

Keep in mind that large systems grow in phases. The Empire Blue Cross system has been in operation since 1988. It is expected to grow to 3000 workstations that will process 100 million pieces of paper per year. The USAA system has been in operation since 1988 following two pilot projects. The reason for the phased approach is twofold: minimize risk and maximize buy-in. Imaging systems will change the way your organization works. You must allow time for this to happen. As a technology, imaging has and will continue to change very quickly. Making excessively large imaging investments today may lock you into a solution that will be quickly outmoded and obsolete tomorrow.

The key to implementing an effective imaging system lies in understanding these evolutionary trends, the maturity of the technology, and the marketplace. In a few years, and certainly by the turn of the century, imaging and all of the ancillary technologies, standards, networks, and economic justifications will be firmly in place in most organizations. Imaging will then no longer be a matter of competitive advantage, it will be an imperative for competitive parity, and a mandate for survival. You must begin to implement and educate your organization today in order to be prepared for that mandate when it arrives.

Rolling Out the Imaging System

When you finally roll out the imaging system, avoid grandiose delusions of making the front page of *ComputerWorld* as IS Professional of the Year. Despite the many showcase examples of imaging success stories, the majority of victories are small, quiet ones. They lead to grander schemes, but setting grand expectations from the outset implies full appreciation on your part of the changes that will take place after introducing imaging. To call that arrogant would be kind, to call it foolish might be more accurate.

Set realizable short-term expectations and apply the technology to a smaller workgroup before establishing a precedent for the enterprise. Introducing imaging to users requires that they invest time to learn a new set of commands, GUI, icons, and potentially a new desktop interface. That is asking a lot. You can go to the well only so many times in any given period of time. Since your objective over the long term is to connect the entire enterprise, your approach should be one that makes sense and has been "field tested" first. To do this your imaging system should be able to traverse the variety of platforms, operating systems, and applications that are used throughout the enterprise. Good imaging should also be as unobtrusive as possible. Support of platform features such as DDE, OLE, Publish Subscribe, aliasing, and multitasking are all important aspects of integration with existing applications.

Finally, remember that the introduction of technologies such as imaging, text retrieval, and workflow can have a dramatic effect on your organization. It is virtually impossible to predict and plan for all the ways in which the introduction of these technologies will alter the ways in which your organization can function. Don't expect the process of redesigning your business to end. Reengineering is neither devastating nor finite. If you plan the reengineering of the business as a finite process, you will be back to where you started in another five years doing it all over again; you will never escape the technology turnstile. Designing a cost-effective imaging system need not be a devastating experience. An incremental approach that stresses education, demonstrated and leverageable results, and user enlistment is best approached as a life-cycle process rather than a once-per-decade decimation of the enterprise. (For a more detailed discussion of phased implementation of EDMS, see Chap. 7.)

With so many options to choose from and the promise of newfound payback with reduced costs, it is difficult to imagine that there could be a downside to this marvelous technology. But cost is only one facet of a cost-effective implementation, and a misleading one at that. You should also focus on the value to be derived from imaging. The higher the value, naturally, the higher the cost-effectiveness. This applies to any level of investment, but is especially important to keep in mind when implementing low-cost solutions, which can create an illusion of productivity while in reality becoming a productivity black hole.

One vendor I spoke with told me of a visit he made to a high-ranking military officer who had just purchased an imaging system for a few thousand dollars. The officer was beaming with pride when he described how he was scanning all of his correspondence on-line with imaging, and it required that he spend only one hour each day scanning in his paper!

Ultimately the myth and reality of imaging are flip sides of the same coin. The myth about imaging is that you can reduce the cost of system installation by circumventing any attempt at business analysis and evaluation of work processes, and merely add imaging to your technology suite. This approach to system installation will not only result in a low payback, but in many cases will leave a sour taste in the mouths of users who participated in the initial application. Use the introduction of imaging, or any EDMS technology for that matter, as an opportunity to evaluate the existing process, enlist users in the analysis, and establish success metrics for the evaluation of success. This will create a long-term partnership with users, which is essential to an imaging system's cost-effectiveness. It may be hard to accept at first, but after all is said and done, it is the users who will define the benefits and payback of the system, not IS.

In reality, low-cost products facilitate the initial acquisition of technology and provide easy installation and a user-friendly front end, but do nothing to change the complexity underlying enterprise business problems. At whatever level you start, it is inevitable that imaging will spread through your organization's information systems. If you view imaging as an intricate part of the business process, and as a simple component in an array of EDMS technologies used as a means of empowering users, you are far more likely to make the right choices and achieve the goal of cost-effective imaging—regardless of the actual products you choose.

Often Asked Questions About the Imaging Industry

Why are organizations investing in imaging, and how have these organizations justified the imaging system?

The most important reason for buying an imaging system is common to all EDMS investments: faster and better access to documents. The ultimate benefit is increased productivity and cost reduction. The long-term result in the commercial sector can also be thought of as competitive advantage. Remember that document management is part of a business process. Many of imaging's benefits result from improvements in the overall business activities which it supports. This is most evident in high-volume transaction systems such as claims processing, customer correspondence, and insurance and loan application processing. High-volume systems generate large savings. USAA estimated savings of $4 million per year from its imaging system; Consolidated Freightways is saving 375,000 worker-hours per year. It is important to keep in mind, however, that maximum payback from an imaging system often comes when imaging is implemented as a component in a multi-technology EDMS, and preceded by a rigorous reengineering of the business processes.

Few if any systems are justified on a singular approach. Most systems are justified with a combination of process improvements, increased productivity, and cost savings. Since many of these costs are not easily evaluated due to their intangible nature, it is possible to justify through demonstration only. Then the question, of course, is how to justify without implementing. An apparent catch-22? Not necessarily. An incremental approach that identifies a key business process and then focuses on one aspect of it that can benefit from being streamlined through imaging may be more than adequate to parlay into a much larger justification. This is precisely the reason for the popularity of the "pilot" approach to most imaging implementations.

How does imaging fit in with other electronic document management technologies?

Imaging represents one component of document management. Its value lies in its ability to store and display a visual representation of the document. It is therefore closely oriented to paper-based documents. This is one of the shortcomings of an image-only solution. One of the areas you should be most concerned with is the ability of the imaging component to work with and compliment other technologies and data types. For example:

- Multimedia support extends the contents and capabilities of the document beyond paper to audio and video.
- Most imaging systems support text fields (e.g., customer name) in the image index database. Text retrieval extends alphanumeric data fields to variable length text fields or text files. More importantly, text retrieval allows sophisticated document access based on document content as well as on indexed fields. Chapter 5 provides a detailed discussion of text retrieval technology.
- Workflow software addresses the direct relationship between the documents and the business processes in which they are used. Workflow software automates work processes by proactively managing documents through automatic scheduling, routing, and tracking. (See Chap. 3 for a more detailed discussion on this technology.)

It is the integration of imaging, text retrieval, multimedia, and workflow technologies that is redefining the document and, in many cases, the business itself.

What are the biggest obstacles to implementing an imaging solution?

The cost of the system is most frequently cited as the biggest obstacle. This is a deterrent for authorizing imaging systems, especially in governmental organizations. However, cost concerns only delay the purchase, since organizations, including the government, believe imaging has clear benefits.

The biggest barrier to implementation is your own organization. This includes the challenge of reengineering the existing IS systems and business process, especially when making the transition from a paper or microform system. MIS and users may resist the proposed EDMS solution during the requirements planning and implementation because of the tremendous change it will bring. The problem lies in the fact that they can easily measure the disruption of the status quo but they have no benchmark by which to measure the benefits of the new environment. To maximize your chances for a successful implementation, pay close attention to the following:

- Educate and involve users, MIS, and management. This will speed approval and implementation.
- Plan for reengineering the process. Focus on making reengineering a benefit instead of an obstacle.
- Formulate a plan for integration with other systems, prior to product selection, if possible.

Questions to Ask When Evaluating an Imaging Solution

While there is no substitute for a solid education in all the technologies that make up an imaging application, the ability to get the right information through intelligent questions is at least half the battle. The following is a list of questions that you should be able to answer with confidence before proceeding with an imaging implementation. These questions force you to focus on the issues that are often left out of an evaluation.

- Has a system audit been done to determine the exact nature of all information (electronic, paper, microform, or other media) that will be accessed by the system?
- How much of this information can or will be converted for access by the system? If you can identify the active portion, there may not be a need to convert the entire backfile.
- What will the backfile conversion costs be? You will probably spend from $.50 to $4.00 per page for conversion depending on a number of factors including document quality, OCR characteristics, compound images, and text.

- Will users need to access the information based on descriptive text through a content-based retrieval functionality?
- Has an organizational sponsor with adequate clout been established for the imaging system?
- Has a sound methodology been used to determine which applications are best suited for imaging, and of these, which one will act as the pilot?
- Have you identified a set of organizational metrics for establishing the means by which the success of the pilot will be measured?
- Which standards are supported by the imaging system? Although there are many, you should look toward existing or proposed standards such as CCITT G3/G4 and ODA for image format and compression, FDDI for networks, TWAIN for peripherals, and ISDN for telecommunications.
- Is an integrated database part of the system or can one be linked easily to the system?
- What are the network requirements for your intended solution?
- What are the legal ramifications of using an imaging system on regulatory, evidentiary, or records management requirements?

Keep in mind that this is a rapidly changing technology that demands a fair amount of ongoing education on your part. Take the time to develop your awareness of the market and technology trends and you will inevitably save much more than time in achieving your goals.

Case Studies

Blood Banks and Health Care

Application requirements:

- Continue current applications and data without disruption.
- Provide diverse platform support.
- Share information with existing DBMS.
- Manage the user's response to change.

The AIDS epidemic has put an enormous burden on health care providers and blood banks. Hundreds of dollars are spent testing every blood donation. The regulation and stringent nature of FDA involvement in the process is such that human blood is now considered, for all practical purposes, a drug. Consider that six separate products are extracted from each of the millions of blood donations received every day. Advances in technology have resulted in up to a 10-year shelf life for these products. If a donor is diagnosed with HIV during that 10-year period, every product created from the donor's blood must be traced and pulled out of the blood supply. The paperwork required to do that is overwhelming. But the process has to be flawless. To further the problem, the information systems on which blood data is tracked must be FDA certified and therefore cannot be altered without significant elapsed time and cost.

The answer was an imaging solution that did not interfere with the existing systems in any way. The unique feature of this solution was the ability to work directly with the organization's information systems without modifying any preexisting programs or code.

Several new vendors have taken this approach through the use of existing system facilities for application integration, as is the case with the use of OLE, or through clever methods of monitoring screen painting and associating alphanumeric screen display items with images. Whatever the specific nature of the integration, it is important to keep in mind that coupling images with current applications can be a valuable means of introducing the technology and its benefits to an organization with minimal disruption. But certain questions should first be considered. Does the integrity of the images and the database need to be synchronized? (If you delete an entry in the database that refers to an image, does the image still exist?) Is the integration seamless or will users be required to jump back and forth from one product to another?

In most cases these are issues that can be easily resolved. You can then proceed to introduce your users to imaging within their zone of comfort before striking chords of fear by proposing radical change.

Regulatory Compliance

Application requirements:

- Fast remote access to images and documents
- End-user capability with little or no training
- Information retrieval based on "fuzzy" criteria
- Minimal initial investment

Since the chemical spill at Bhopal, India, the chemical industry, the *Environmental Protection Agency* (EPA), and the *Occupational Safety and Health Administration* (OSHA) have increased their emphasis on the accessibility of safety procedures and information relating to hazardous materials. This requires the ability to quickly retrieve information about plant facilities and materials that may be required to respond to any incident involving hazardous materials.

Much of this information is contained in vast repositories of engineering drawings, plant maps, materials safety data sheets (MSDS), and a myriad of regulatory guidelines. Access to this information has historically been through a number of interfaces, including paper, blueprints, regulatory guides and updates, company memoranda, maintenance logs, and a variety of on-line databases.

OSHA has, however, put pressure on these organizations to put all necessary documents and data in electronic form in order to provide immediate access to critical information. The problems this creates are monumental. Previously separate systems must be integrated and coordinated into a single EDMS with comprehensive access, retrieval, and routing functions. There are few products that can meet such a broad range of demands. Organizations have to turn to integrators capable of bringing together the individual technologies and products required.

One example of an integrated approach to this type of application is the novel and cost-effective approach employed by Lockheed Advanced Development Company. It did not require engineering-specific functions. Lockheed chose an integrated imaging, OCR, full-text retrieval package on DEC VAX and PC workstations. The novelty of Lockheed's approach was a dial-in and fax-back capability that allowed remote access to all documents stored on a central image server. Field personnel could dial into the imaging system using a standard Touch-Tone phone, request an MSDS by following a simple dialogue with the automated system, and then receive it at a fax number they specify. The system has the added capability of retrieving documents based on the textual content of the document—even if they are not able to precisely identify the term or word they are searching for. This is accomplished through a process known as a *fuzzy* search that retrieves documents based on a neural network approach to word indexing and retrieval. (For a detailed explanation of fuzzy searching, see Chap. 5.)

Even on a small collection of 26,000 documents the system initially resulted in savings that exceeded $250,000 per year.

Cost-Effective Imaging in the Courtroom

- Scanning and OCR
- LAN-based operation
- PC-based platform support
- Minimal integration
- Bundling with text retrieval, OCR, compression and decompression
- Client/server architecture

Drug trafficking results in an immense amount of money laundering. This chaotic transferring of funds from bank to bank and business to business leaves behind a monumental paper trail. It is the job of a joint IRS/DOJ/DEA task force to prosecute money laundering cases. That means shifting through mountains of paper that include testimony, financial documents, deposit slips, check stubs, endorsements, and an unending variety of paper shapes and sizes.

Recently a pilot was undertaken to assess the viability of converting trial documents (all of the paper used for a court trial) into an imaging system. With a low budget and little in-house expertise the solution needed to be not only inexpensive but easy to install and use. The product chosen was a PC-based solution that supported client/server architecture and a Novell Network, and provided integrated OCR, full-text indexing, and peripheral support for accelerated scanners and laser printers. The cost for a five-user license including hardware, software, and integration was slightly over $100,000. The high bid received by the same user for another integrated solution was almost $900,000—of which $600,000 was the cost for integration services! Luckily they did not have the budget and decided to look elsewhere.

The benefits of the approach they chose were its compatibility with existing hardware and networks and a virtual turnkey image-applications interface. The system also included a full-text retrieval function that was automatically enabled for every document scanned as an image. This allowed users to retrieve a document based on key fields or any text that was part of the original document.

The system is intended to provide both the prosecuting and defending counsel with trial document on optical disc, eliminating the enormous task of copying and shipping original documents back and forth. It may also find its way into the courtroom for use during a trial.

Imaging: The Technology

Types of Imaging Products

Imaging products can be grouped into a variety of categories based on pricing, platforms, networks, interfaces, and many other considerations. The difficulty with these categories is that they do not fit neatly into the classic single-user, workgroup, enterprise model that we have come to know. A better means of categorizing imaging systems is to use a two-dimensional matrix model that measures the key differentiators of every imaging system's customizability and scalability. No two factors are as important to understand since the two things that can be said about any successful imaging system are that it is bound to change and it is bound to grow.

The ability to customize a system is represented in products from off-the-shelf to tools-based imaging. Scalability can run from single-user to enterprise-capable. On a matrix of these two categories we can plot each imaging product into a quadrant that represents its primary position and capabilities. As Fig. 4-4 shows, there is significant overlap between the categories but you will not find any one product that effectively spans the entire spectrum in terms of both functionality and performance.

There are also distinct price breaks along both categories, which make it impossible to say that one product is cost-effective across all

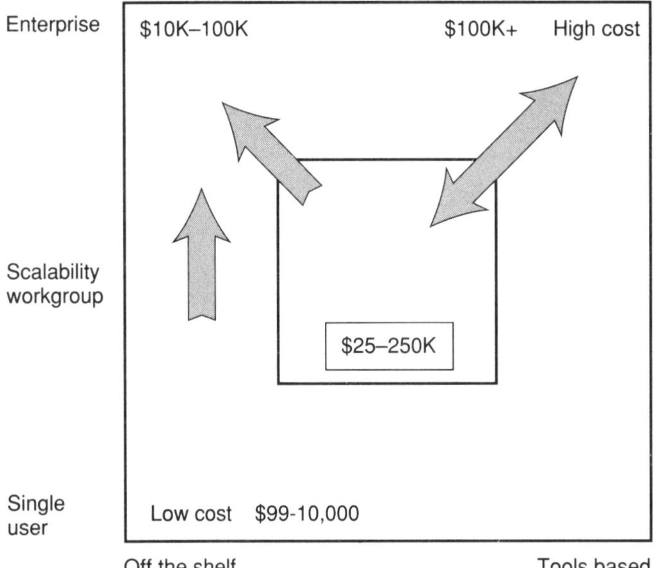

Figure 4-4 Four-quadrant imaging matrix. It is difficult to identify many imaging products by exclusive categories. This chart shows the price range and fit of popular products based on their scalability and the extent of possible customization. Products in the lower left of the graph have a low-cost entry price but can often scale into enterprise solutions if a common GUI such as Windows is used throughout the enterprise and customization is not a requirement. Products in the upper right often cost more but allow for extensive customization and integration with existing applications and large numbers of users. Those products in the middle box provide a wide range of scalability options and customization tools. Their cost can, however, quickly mount in an enterprise application with a large number of users. The arrows show market movement for each of the vendor categories. It should be noted that many of the low-cost solutions may encounter problems with high-throughput transaction-based applications.

applications. Matching the characteristics of these product categories with your requirements is a significant first step toward identifying candidate cost-effective imaging solutions. The imaging products to keep a close eye on fall into an area that forms a union of each of the four imaging categories. These products provide scalability and customizability at a moderate price. As the imaging market matures this category will represent an increasing part of the market for imaging systems, drawing into it both high-end and low-end vendors. Let's look at each category clockwise from the lower left to the middle. The lower right is not addressed since tools-based applications are built for larger workgroups and not single-user environments. If you happen to be a single user with a strong desire to custom-build an imaging application and the budget to match, you could of course purchase an expensive workgroup or enterprise application with which to do this.

Most of the confusion today rests with products that occupy the lower-left quadrant of the matrix. The greatest variety of solutions and the largest number of new product entries are in this space. These not only represent inexpensive imaging systems but in many cases include basic OCR, fax, scanning, workflow, and text-retrieval functionality. Products of this type can be beneficial for introducing users and workgroups to imaging at a low cost. But the tools provided to build applications on them are minimal and may lead to a hard stop in short order as user demands for customized applications and interfaces increase.

At the high end, products are highly customizable but also relatively expensive. These products are most cost-effective for applications with large numbers of users that demand integration of multiple applications and technologies. Significant process redesign and customization of interface components are typical in these installations.

Tools-based products offer an ideal development platform for *value added resellers* (VARs) who repackage the technology with vertical emphasis for specific applications. These are not found as often in enterprise environments today. This is not so much a reflection of the technology as it is symptomatic of the desktop platform's minimal role in enterprise applications. In addition these systems do not have the traditional track record in enterprise information systems of major vendors.

Enterprise/off-the-shelf is a category that many would claim to be an oxymoron. But the advent and strong appeal of Windows coupled with the broad use of networks and E-mail have established an infrastructure and a precedent for this type of imaging. The availability of a standard GUI such as Windows is an absolute must for an enterprise off-the-shelf solution. Windows provides a common means of access and integration with other applications. Products strongest in this category are those which take full advantage of Windows tools such as DDE or OLE and have support for popular networks. What is not available in many of these solutions is the ability to tightly integrate imaging with existing applications and customized user interfaces through the availability of a 4GL or extensive API library.

This, however, is where the greatest area of debate can be found in discussing cost-effective imaging. There are at least three distinct levels of integration of imaging with an information system. Think of this as a third dimension of our grid.

The first level of integration, the *graphical user interface* (GUI), provides basic hooks into other applications. This may be accomplished through a product that uses the standard conventions of a Windows interface to represent file folders, click and drag, and other familiar metaphors. The approach is appropriate for basic imaging requirements. The payback of these applications is in the elimination of paper, faster access to documents, and the transfer of images through a messaging or workflow process. OCR, scanning, and text retrieval may all be part of the package as well. Other applications, such as a spreadsheet or a word processor, can be tied into the user's imaging environment through the use of facilities that launch the application directly from the image interface.

The second level of integration links the imaging system to other applications through callable modules. Examples are the ability to invoke an imaging system through an *application program interface* (API), invoking it through a Windows facility such as OLE, or simply associating it with screen display areas (often called screen scraping). These modules offer the full functionality of imaging without requiring extensive programming or integration. Many of these packages use proprietary database engines to perform indexing and retrieval. This can create problems when migrating to a more advanced imaging system or trying to integrate imaging systems across the enterprise.

With both the first and second integration levels, the interface components of the imaging systems are relatively standardized. This can be a benefit to quick application development and user acceptance, but it can also present a problem for sophisticated applications or especially novice users who are not accustomed to a specific GUI environment. In these instances, developers need to take a significant step forward to imaging systems that offer open approaches to integration and customization. These products often support industry standard database engines and 4GLs. Their primary benefits are highly customizable configuration and interfaces, integration capabilities with existing applications, and programmable extensions such as workflow routing modules.

The Components of Imaging

Imaging is a straightforward technology. Despite that, all too often evaluators get caught up in the underlying details, which can easily become intimidating. A basic understanding of imaging, sufficient for the vast majority of applications and implementation, is not tedious or difficult. We can begin establishing this by looking at the components of imaging. Every imaging system, small and large, consists of the same basic array of six components:

- Capture/scanning
- Indexing
- Storage
- Retrieval
- Workflow/routing
- Presentation

In short, imaging will take an existing source of information such as a picture, a three-dimensional object, or a page of text and convert it into an electronic format using a scanning device that takes the analog information, digitizes it, and creates a computer-based binary representation. The resulting electronic image is indexed for retrieval and filed in an on-line storage device. Whether it is optical or magnetic makes no difference in the process. A search and retrieval engine will be used to recover the image. Finally a display or an output mechanism will present the image. In its most basic terms, that's all there is to imaging.

Capture or Scanning. This is the conversion of existing paper-based information (documents) into electronic form (images). The process may include OCR, which will convert all or part of the textual portions within the scanned document into machine-readable form, such as an ASCII text file or a word processing file. (For a complete discussion on OCR processing see the OCR section in this chapter.)

The capture component of most imaging systems is represented by the physical processing of documents through a mechanical scanner. Although seemingly a trivial process, there are numerous issues that should be considered for even a small workgroup application. These include support for scanner drivers, throughput rates, double-sided scanning, OCR accuracy and speed, compression and image formats used, forms removal, and document quality. Each can have a significant effect on the cost-effectiveness of an imaging solution. Fortunately however, a broad range of scanner features and functionality are available across solutions, from low end to enterprise wide.

For example, image compression and decompression are supported through a variety of approaches. At one time, the only viable approach to this issue in cases of demanding high-volume imaging applications was the addition of hardware for the server and the client workstations. Today, software-enabled compression and decompression are not only practical, but preferable in many cases where the host processor is a 486 Intel-based or RISC architecture. Software is easier to upgrade, which is an important concern, considering the rate at which new compression algorithms are introduced. For instance, *Joint Photographic Experts Group* (JPEG), which can achieve up to 100-to-1 compression ratios, is an ISO standard for applications using color images or photographs. But in the next few years, it is likely that newer compression standards such as *Fractal Image Format* (FIF) will increase the compression ratio by another order of magnitude. Keeping up with these changes will be an ongoing expense that you must consider, and one that is facilitated through software-based compression.

Indexing. Indexing may be one of the most often overlooked elements of an imaging system. In the case of indexing there are distinct differences among low-end, off-the-shelf products and high-end customizable solutions. The primary difference is the support of a standard database engine as opposed to a proprietary database or filing system. Although both approaches may provide the functionality needed to index and retrieve the documents in the initial application, a proprietary DBMS will create havoc when migrating to another imaging system or integrating multiple systems across the enterprise.

There is another dimension to indexing that is not as easily found in most products. That is the ability to index a document on its complete textual content—often referred to as full-text retrieval. Full-text retrieval offers a cost-effective solution and a great deal of flexibility for users of document collections rich in text content. Full-text can be provided either through the integration of a third-party product or a proprietary offering bundled with the imaging vendor's product. These systems must, of course, also provide an OCR component in order to convert the scanned image to text. (See Chap. 5 on text retrieval for a full discussion of this technology component of EDMS.)

Storage. Storage is the component most often left to the user or developer to choose. Few solutions provide a storage mechanism as part of the imaging system; however, the absence of a bundled offering should not present a problem. Although the storage overhead of a large imaging system would make it logical to use a form of optical storage, optical storage is not a requirement of imaging. In cases where it is determined that an optical storage facility will be used, implementation is still fairly straightforward because optical storage technology is easily integrated with any system that supports an SCSI (Small Computer Systems Interface). Problems may come into play, however, when you begin to deliver imaging over a client/server architecture. Here the support of a particular network can become a critical issue. (For a complete discussion of the practicality and implementation issues related to optical storage, see the Optical Disc Technology section in this chapter.)

The intelligence of the imaging system's storage configuration is also something to consider. For example, how are jukebox operations handled? Is caching to disk or memory provided? What network services are used to avoid bottlenecks and long queues? Although you may start with a small system, be sure to ask these questions if you have any plans to migrate to a larger implementation.

Retrieval. Retrieval is innately related to the indexing scheme used. A filing system approach, found typically in smaller, single-user systems, will use a filing cabinet, folder, and document-name metaphor to retrieve images. The DBMS approach, which is the most popular available, will provide key field retrieval for a set number of items that are entered manually by a data entry operator during the scanning process or are extracted by a field-oriented OCR. A text retrieval approach will provide retrieval based on the textual content of the documents. Of course, the text retrieval approach mandates that the scanning process includes OCR processing.

Distribution. The most popular approach to distribution of images uses the file system on a client/server architecture to allow for distributed access to the images. In a client/server model, the images exist in compressed format on the image server and are sent to the client for decompression and viewing. The images do not permanently reside on the client workstation. This is especially important to adhere to since images can require significant storage space. A single-page compressed document, for example, will take up to 50K, 10 times the capacity of a similar page of text; uncompressed, it increases tenfold to 500K.

The integration of workflow software should be considered when evaluating the distribution requirements of an imaging system. Workflow does not require that the end user "request" an image, but rather proactively routes images throughout the network based on associated work process rules. Workflow is being laid claim to by virtually all image vendors. It has, unfortunately, become an ambiguous term because of its widespread usage. True workflow automation requires a sophisticated set of tools for the definition of routing rules and the reporting of document activity and usage. In reality, only a few vendors actually provide this level of functionality. (See Chap. 3 for a complete discussion on this technology component.)

Display or Printing. From the user's standpoint, display and printing are the most important considerations of all for production imaging systems. Unfortunately many new users of imaging have an expectation that the desktop metaphor used with data entry screens and information systems designed for a small screen area are appropriate for an imaging system as long as the resolution of the document is from 120 DPI and higher. Nothing could be further from the truth. Since most document imaging systems deal with documents that have a standard page layout, users of a standard monitor are saddled with the burden of scrolling, panning, and zooming through interminable iterations to view multi-page documents. Despite advances in the technology such as scale to gray software that enhances the display of black and white images, screen size is still a key factor in the ergonomics of production imaging solutions. A cost-effective approach may use a system that supports scale to gray for VGA monitors but also has the capability to work with larger portrait and dual-page formats.

Finally, don't dismiss the importance of paper. Promises of paperless offices aside, paper output is still a cost-effective means of document transfer because of its intrinsic portability. Therefore, in applications where it is likely that output will be produced in paper to one degree or another, it is important that the printer specifications are carefully considered. Relying on standard laser printers that have not been accelerated to accommodate large image files can be a frustrating experience at best. Consider upgrading to a departmental printer or using an imaging accelerator for your printing device. Some vendors provide added intelligence within their imaging product to speed the printing of images. Most offer accelerator boards from third-party manufacturers to accelerate the printing process.

Imaging Architecture

Naturally there are alternatives in each of these components, but before we get into these, let's talk about the alternative architectures for imaging solutions and their associated costs. There are three fundamental imaging architectures: stand-alone, host-based, and client/server.

Stand-Alone Systems. Think of stand-alone imaging as an electronic file cabinet. Its purpose is simply to replace paper with electronic storage. Stand-alone imaging systems are most often used for very specific imaging applications that rely on large volumes of documents for reference purposes. The stand-alone system may be used as either an image server, providing images to a network of other desktop or host platforms, or it could be dedicated to one workstation or LAN. In either case, it replaces the file cabinet as a repository for documents. In some cases, that may mean a relatively closed system with minimal facilities for integration with external applications. Most imaging vendors can support a stand-alone environment. The question is not one of availability, but rather application. Why implement a stand-alone system? Admittedly there are few applications where a stand-alone imaging system will not need to integrate with other information systems and applications through an enterprise.

A stand-alone imaging system consists of all the basic components required to perform imaging, including a host system that manages the image processing, backup facilities, servers for scanning documents, printing documents, workstations for each user, and one final component that cannot be ignored—the network.

Host-Based Systems. The host-based environment is generally applied when an existing corporate repository of documents for use across an enterprise is planned. Most often this represents a transaction-oriented application that is augmented with imaging. The host-based architecture relies on a mainframe or minicomputer as the image repository, server, and manager. Intelligent devices such as X terminals or workstations will be required to enable image capability at the desktop. All remaining components of scanning and printing are performed through the host. Additionally, compression and decompression hardware or software will be required to enable efficient image transmission. Although this sounds fairly simple it can be an expensive solution for small workgroup implementation due to the high buy-in level of the host-based imaging software.

Client-Server Systems. The client-server environment is the predominant and preferred imaging architecture. Client-server imaging provides processing intelligence at both ends—the client and the server. In this architecture, the server acts as one of several potential providers or repositories of the images and another node, the client, acts as the requester of the images. But the provider–requester relationship alone does not establish a truly client-server imaging environment. Simply being able to download an image to the client does not establish client-server. The client must have some intelligence or processing power by which it can work with that image.

In this scenario a client could also act as a server on occasion. For instance, a LAN server may act as the repository for certain images in a departmental network, while simultaneously another local node may act as the repository for images that are proprietary to one workgroup within that same department. Client-server applications can become very complex as a result of these types of relationships. It is wise to be careful when you evaluate a vendor based on their ability to work in a client-server environment. Be careful to ask the right questions, such as whether they simply download information or whether they have the ability at the client to work within the entire range of the document management functionality. If the vendor indicates that the client acts solely as a recipient of downloaded images, this is a clear indication that the imaging system's flexibility and responsiveness are restricted.

Input Mechanisms

Scanner Technology. The scanning and OCR components of an imaging system digitize an analog image. OCR, which is often confused with scanning, is a secondary and separate step that processes the textual portions of the scanned document. Since many scanners come bundled with OCR software or hardware the two are usually considered a single process and technology, which is not the case. When determining system requirements, it is important that you appreciate the differences between these technologies as well as their relationship.

The scanning process is straightforward. (See Fig. 4-5.) In basic terms, a paper document is recognized by reflecting a light source off the document and recording the intensity of the reflected light through a series of *charge coupled devices* (CCDs). CCDs are light-sensitive analog instruments that provide continuous readings of light intensity. The scanner will take this analog signal, perform an analog-to-digital conversion, and compile these readings for the entire document. Once this information has been compiled it is compressed.

The digitized data is represented in DPI. For example, a typical business document is usually scanned in at 300 DPI. There are several critical issues to be aware of in this regard. First, that same "typical business document" can be scanned at anywhere from 100 DPI to 800 DPI. Why the differences? In reality, there may not necessarily be a difference. A 300-DPI document and a 600-DPI document may contain the same amount of information gathered from the original scan. The difference may simply be that the scanning process interpolates sampled light readings to determine the effective 600-DPI resolution. In other words, it will create three-fourths of the information in the document from the actual sampled data.[4] When evaluating scanners, both the sampling rate and the effective DPI should therefore be considered.

Between 300 and 600 DPI, the difference can be barely perceived by an untrained eye without magnification. Beyond 600 DPI, it becomes exceptionally difficult for you to discern a difference in a bitonal

[4]Although the DPI only doubles from 300 to 600, the total number of "dots" required if DPI doubles is four times the original DPI (300 × 300 = 90,000 dots per square inch; 600 × 600 = 360,000 dots per square inch; 360,000/90,000 = 4).

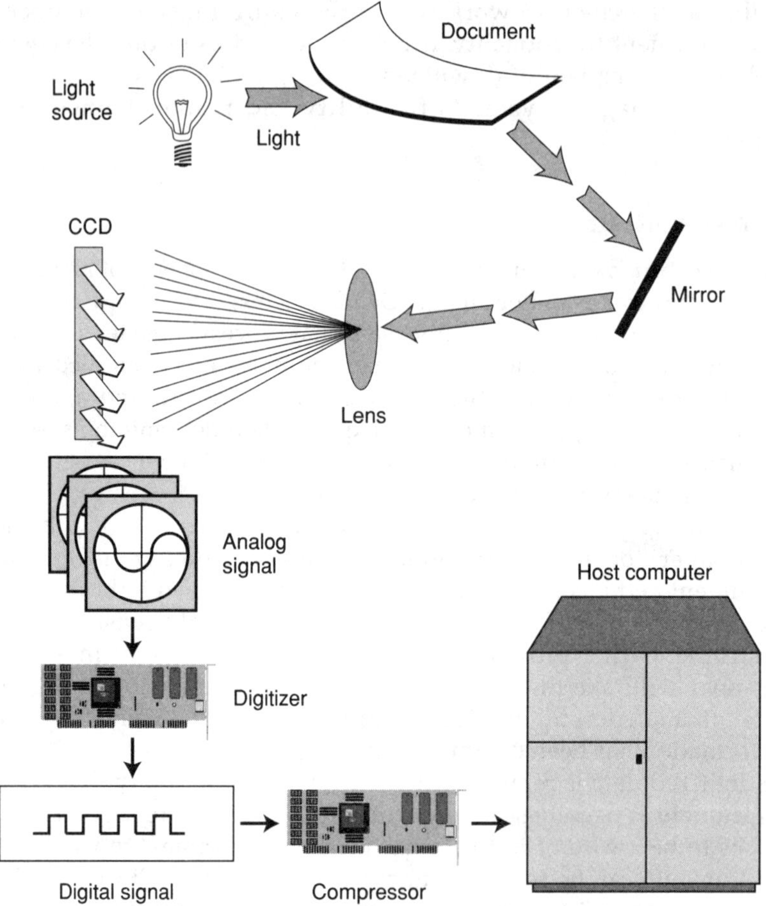

Figure 4-5 Scanner technology schematic. The principles behind scanner technology are relatively simple. A light source reflected off a document results in patterns of light and dark which correspond to white and black shades on the document. The light intensity, which is recorded by light-sensitive CCDs, determines the presence of information which is then recorded as a bit map of the document (a matrix of 1s and 0s). This bit map is then compressed to reduce its size before being stored as an image file.

image. What you will discern is an overall increase in the quality of the document as the DPI increases. But it becomes very difficult for the human eye to discern any specific difference in the document itself without magnification.

As a result, 600 DPI is the upper range for a business application. Generally, 300 DPI is sufficient for business documents, including those images that will be processed by OCR. If the business documents are not going to be searched using a content-based application, then the conversion of image to machine-readable text will not be necessary. As a result, OCR processing will not be necessary and a lower DPI can be considered. (See the OCR section in this chapter for more details on OCR technology and requirements.)

Keep in mind that when you scan documents you are creating and investing in an information asset base. You need to be very careful to preserve that asset. The last thing you want to do is scan the documents and then have the users' requirements change in such a way that rescanning will be required. Scanning should be a one-and-done process. Anticipate future imaging requirements, and scan to the high end. For example, if OCR will not be performed on the documents as part of the initial system implementation but it is anticipated that OCR will be performed at a later date, then, despite current requirements, the documents should be scanned at 300 DPI (or higher) to accommodate the OCR needs in the future.

Scanning Compression Image Storage Formats. The bit map or image file created for a 300-DPI document contains more than 8 million separate bits of information.[5] That's a large data file by any measure but especially when you consider data transmission over standard networks. As you increase the depth of the document in bits by adding gray scale or color to the scan, the size of the document will increase significantly. In fact, at 400 DPI a document scanned at 256 levels of color or gray will translate into an image file that is upward of 15 MB in size. The size of image files is an obvious concern, in terms of storage capacity and transmission speed. Therefore, file compression is an absolute necessity for imaging applications. While every imaging solution will undoubtedly provide compression, you must be aware that there are several approaches to compression, making for another area in which design decisions will have to be made.

Types of Compression. There are basically two separate categories of compression standards: *lossless* and *lossy*. Lossless compression algorithms do not destroy or lose any of the original data from the scanned image file. Compression occurs from the elimination of data redundancy in the image file; however, the elimination is handled in such a way that the original data values can be recalled in their entirety during image decompression or expansion. (It is important to remember that when used in the context of compression, lossless means no loss of information

[5]Calculation of storage requirements based on DPI and size of page ($300 \times 300 \times 8.5 \times 11 = 8{,}415{,}000$).

from the scanned image itself. Loss of information will occur from the scan process itself. For example, if the original document is a continuous-tone image, any digitization of the document will result in lost information.[6] A 300-DPI scan captures only 90,000 bits of information for each square inch of a document, regardless of how much information may actually be stored in the document.)

Lossy compression however, does eliminate some of the information from the scanned image file. You may start off with a 300-DPI document with 8 million bits of information, but may end up with a document that has 8000 pieces of information. Lossy compression literally eliminates information that has been judged to be superfluous. When the image is reconstructed the pieces of information eliminated during the compression process will not be recallable; instead, their values will be approximated through a variety of algorithms.

Lossless. The most popular lossless varieties of compression used in imaging are CCITT[7] Group 3 and Group 4. Virtually every imaging vendor supports at least one of these versions of the CCITT standard. Both are ISO standards and provide reasonable compression. As shown in Fig. 4-6, CCITT standards provide varied degrees of compression depending on the type of document being compressed. Since CCITT is an international standard, it provides transportability across imaging systems and reconstruction of the image without losing any of the information that was in the original scan.

The basic process of compression relies on the underlying redundancy in a document. In fact, all compression methods whether lossless or lossy rely on this fundamental principle. In the case of CCITT, the redundancy exists in what is referred to as *run-length redundancy,* any contiguous series of black or white dots on a single line of a document scan. The compression methodology will store in a piece of information to record the fact that there is a run length of *n* length of a certain value (black or white), rather than storing the color value for each individual unit within the run length. Group 3 compression will track and eliminate this redundancy across lines on a scanned image.[8] Group 4 tracks redundancy across the page as well as moving downward on the page. The process efficiency will increase as the amount of redundancy increases. Therefore a standard letter document allows for very high compression ratios since it contains a high percentage of

[6]A digital representation, by definition, cannot preserve all of the information in an analog image.

[7]Consultative Committee for International Telephony and Telegraphy (CCITT) is part of the International Telecommunication Union founded in 1865 and located in Geneva.

[8]Group 3 actually eliminates some redundancy from top to bottom of a page by using the Group 4 approach on small sets of horizontal lines.

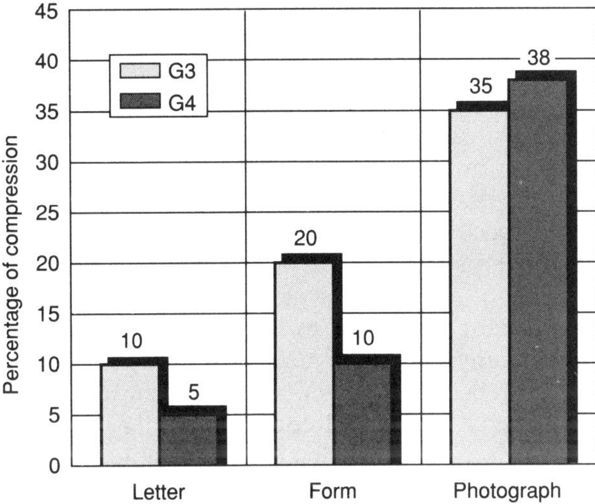

Figure 4-6 CCITT compression.

white space, whereas a form with many more transitions from black to white space will yield a lower compression ratio. A typical compression ratio for CCITT Group 3 is 10 to 1, so that a 500-KB uncompressed image will result in a 50-KB compressed file. Because the information stored in the compressed file tracks the length of each run length and the tonal value for the run length (black or white), the original data values for each pixel can be recalled during expansion of the image file; thus, a lossless form of compression.

Lossy. Like their lossless siblings, lossy compression algorithms also rely on the elimination of redundancy. The difference, however, is that in a lossy method, each bit within a collection of bits (the collection is determined by the algorithm) is regarded as having a value equal to all other bits in the collection. For example, in the case of JPEG,[9] the most popular form of lossy compression, matrices of 8 by 8 bits are quantitized using

[9]Joint Photographic Experts Group (JPEG) is an ISO standard. Unlike Group 3 and Group 4, it requires hardware accelerated compression for real-time use.

algorithms that result in only two values for each matrix. These values represent the chrominance (color) and luminance (intensity).[10] When the image is reconstructed, these average[11] values are used as the values for each bit within the original 8 by 8 matrix. Thus, bit by bit, information is lost (lossy compression), which results in an approximation of the original image. This approximation is often indistinguishable from the original, especially in cases where the original is an information-rich data source such as a color image. Typically, the ultimate processor, the human eye, cannot discern minor variations in color and brightness.

By eliminating much of the original data, a higher compression ratio can be achieved with no distinguishable loss in quality to the reader, despite the loss of data. JPEG does, however, allow for increased compression ratios that will result in a degradation of the image. The user can specify a higher compression by indicating that the image quality need not be optimized in the compression process. Advanced methods of JPEG will also allow the user to specify that different regions of the image be compressed at varied levels of quality. This can be a valuable feature if the image contains large regions of similar color or shading. JPEG typically results in compression ratios ranging from 30 to 1 to 100 to 1.

Another means of lossy compression is fractal-based[12] compression. The mechanics of fractals can become rather complex. In concept, however, it is simple to understand. Fractal compression is based on redundancy. The best way to think of a fractal is to consider that certain patterns of an image are more likely to represent other patterns within that same image than they are to represent patterns within other images. Fractal compression identifies these patterns throughout an image. The computational demands of fractal compression result in its greatest liability—asymmetrical compression. Asymmetrical compression refers to the fact that there are two processes involved in compressing a file and a time associated with each: the time it takes to compress a file and the time required to expand or decompress the file for viewing. In an asymmetrical approach such as fractal compression, the time to compress a file is significantly higher than the time required to expand the file. That difference can be as high as 10 to 1. Fractal image compression is not widely used in commercial products, but it is anticipated that it will become more commonplace as the reliance on images that are excessively large, data-rich files increases.

[10]Quantization is similar to the process of separating a color television signal into black and white, and color.

[11]Average value of a JPEG matrix is not an accurate representation but sufficient for our discussion.

[12]Developed by Iterated Systems, Inc., Norcross, Ga.

Compound Document Scanners. The discussion thus far has viewed the business document as a homogeneous data source. That is to say, we have assumed that the document is either completely text or completely image. In reality, very few documents contain just image or just text. Most business documents contain a combination of graphics, text, and images—all on a single page. The issues in scanning such a document are not just capturing, compressing, formatting, and indexing the image, the means by which you deal with the compound nature of that document . Specifically, how do you preserve a compound document's rich presentation style and its rich information value in the process of scanning, while maximizing process and storage efficiency?

There are several ways this can be done. One way is to scan the document as an image. But that doesn't provide anything more than a static "picture" of the document, along with all of the problems of indexing, storage requirements, and reuse in other document applications. This approach does not take into consideration the various components of the document and exploit the properties and capabilities of each. It reduces the entire document to the least sophisticated data type, the bit-mapped image. For example, images cannot be modified as can a word processing document. To maximize the flexibility and responsiveness of the system and minimize the storage requirements when working with compound document scans, you need to be able to intelligently separate text and image and then reconstruct the document, preserving the visual attributes of the image and the retrieval and format attributes of the text. (See Figs. 4-7 and 4-8.)

Caution should be used in these applications. Although it is easy to scan a compound document and separate blocks of text from images, it must be realized that much of the scanning process and the reconstruction process will be manual. This is not trivial. Those who perform the scanning and reconstruction must understand the makeup of the documents and have a basic awareness of the relationships between the objects that compose the documents.

In applications where extensive reconstruction is required a tenfold increase in the cost of document conversion costs is not unusual. That may translate into a $10 per page cost, or higher, for the conversion of a legacy document collection. In fairly controlled applications (i.e., a small number of document types, similar in construction), this problem can be augmented through the use of a compound document scanner which can automatically recognize the different elements of a document.

In either case, if the scanned images are to be decomposed into separate components, a tagging schema such as SGML, a linking approach such as hypertext, or a database must be used to monitor and manage the relationships between the separate data components. While the

Use templates or intelligent recognition
to separate images from text

Need to:
- Remove forms from images
- Differentiate between graphics and text
- Align, deskew, and stitch
- Provide mechanism for reconstruction

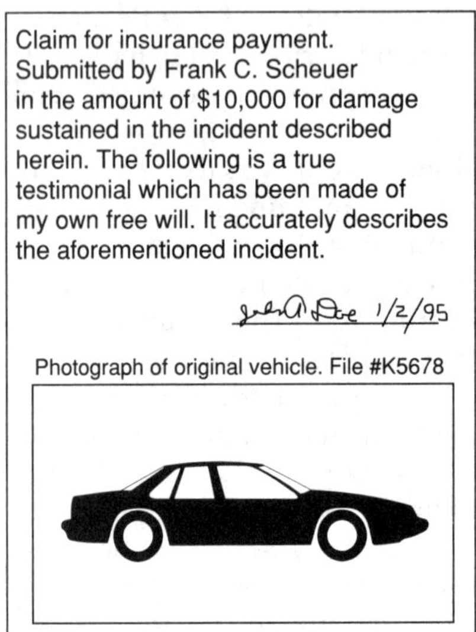

Claim for insurance payment.
Submitted by Frank C. Scheuer
in the amount of $10,000 for damage
sustained in the incident described
herein. The following is a true
testimonial which has been made of
my own free will. It accurately describes
the aforementioned incident.

1/2/95

Photograph of original vehicle. File #K5678

Figure 4-7 Compound document and form processing.

creation of these links, tags, or database values can be somewhat automated, it must again be stressed that even in the best cases a certain degree of manual intervention must be applied to ensure process and data integrity.

Optical Character Recognition Technology

A key part of document scanning is optical character recognition (OCR). OCR is an add-on technology to a document scanner. OCR is the means by which the text objects within a document are converted from a bit-mapped image to a text representation such as ASCII. The scanner decomposes the document into picture elements called pixels. Those groups of pixels which form alphanumeric characters are "read" by the OCR software, and translated into the proper ASCII (or other machine-readable code) character.

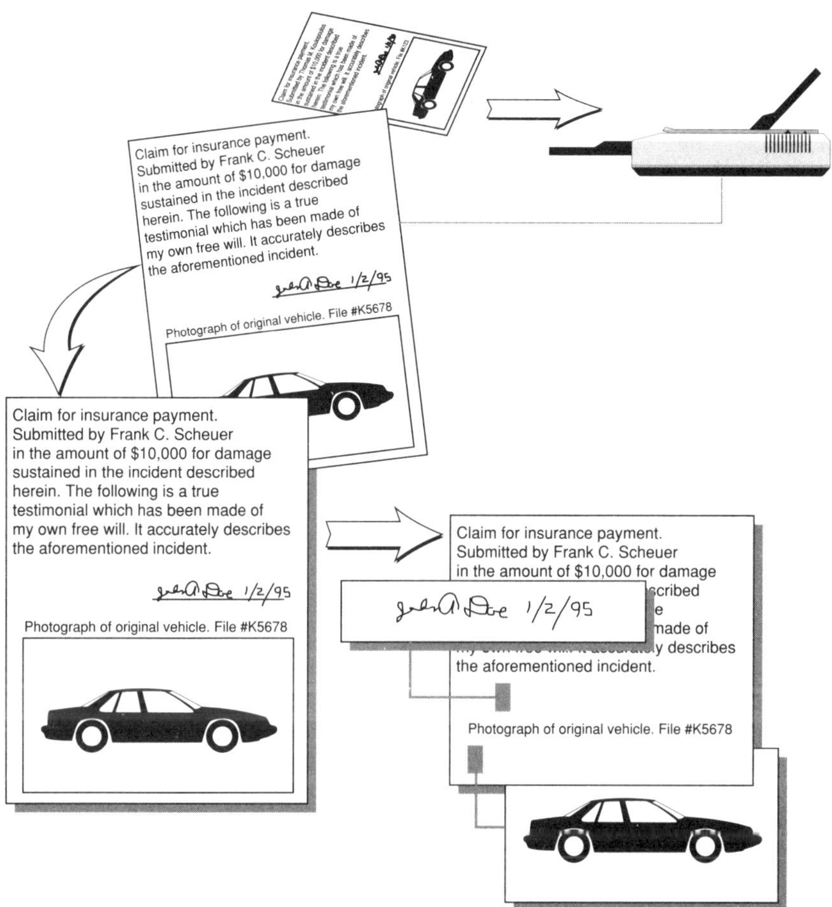

Figure 4-8 Compound document and form processing. Since many documents contain both image and text information, multiple files may have to be created for each page. Compound document scanners accomplish this with a single pass of the document. Note that the removal of lines from a form can result in a savings of 10–20 percent of storage space. Removing all but the entered information typically results in savings of up to 90 percent! If it is necessary to view the original document and space savings are still desired, the document form can be saved once; the entered information on each form can be extracted and saved separately. At viewing, the extracted data from each document can be merged with the saved "master" form.

Until the early 1970s, OCR could support only a limited number of popular typewriter fonts. The introduction of systems that could be taught different fonts as necessary broadened the appeal of OCR to the desktop publishing and text retrieval markets. The technology radically changed in the late 1980s with the development of sophisticated recognition features that allowed OCR to go beyond simple font character matching. Topological analysis has replaced the first OCR technique, template-matching to become the most commonly used technology, allowing recognition of virtually all characters based on their unique features. These new technologies comprise a new set of recognition products, commonly referred to as intelligent character recognition (ICR).

OCR is accomplished with either specialized software or a broad-based dedicated processing system that connects directly to a scanner. Typically, the high-end OCR products provide scanners with OCR based on customized chips built into an external processor. These systems are more expensive (approximately $20,000 and up), but provide higher throughput rates and allow recognition to be done in the background, without tying up the workstation. Low cost and flexibility have made the software-based system popular, especially in applications where input requirements do not warrant high throughput speeds. Software-based products also allow for simple and inexpensive upgrades as new technology becomes available. These OCR packages typically reside on a PC or workstation and work with various scanners.

There are two basic methodologies or approaches to performing character recognition: *template-matching* and *topological analysis.* Though discussed separately, both approaches, or features from each, are often combined into a single OCR product. Topological analysis can also be separated into several subcategories, based on varying levels of sophistication and types of technologies used.

Template-Matching. Template-matching, also known as font-matching or matrix-matching, is the original and simpler of the approaches. The individual characters of the text being processed are matched to existing character templates stored in the system. This approach can recognize only those fonts already known to the system. Products using this approach can store anywhere from a single font (monofont machines) to several fonts (multifont machines). Font-matching is faster than topological-based OCR, but is more rigid. In addition to supporting only the established fonts (including type size), the nature of template-matching requires clean input data. There is little tolerance for any flaws in the text, such as ink smears, faded or blurred text, or miscellaneous smudges or markings. None of the currently commercial OCR products rely solely on template-matching, although almost all use it to perform a first pass against the document.

Topological Analysis

Pattern Recognition or Feature Extraction. Pattern recognition, also called feature extraction, does not utilize templates. It takes a rule-based approach to OCR. Rules concerning the construction of characters are coupled with image processing algorithms that can detect the edges of printed characters. Scanned images of the text are broken into individual elements. The parts of the character and their intersections are compared to a set of rules regarding character construction. For example, an uppercase *D* may be defined as a vertical line intersected at both ends by a half circle. A lowercase *d* may be defined as a half circle intersected by a vertical line approximately twice the height of the circle. This approach can handle multiple fonts and styles such as bold, italic, and pitch changes without requiring a template or a separate set of rules for each font type. Typically, systems using this approach are slower than font-matching systems. Though some text flaws such as ink smears and miscellaneous smudges or markings are ignored, other text flaws and features like broken type, underlining, and overlapping characters are still obstacles to the OCR process.

Omnifont. An advanced form of feature extraction, omnifont systems are totally font and style independent. They use several text handling algorithms known as "experts." These experts are integrated hierarchically by level of expertise and function. Multiple experts, including topological experts for character features such as loop, concavity, and line segment detectors, are combined to perform sophisticated feature recognition. Other experts act as managers that combine the results of both high-level and low-level recognition experts. These systems automatically learn fonts and print styles as they process the text. High-level experts teach low-level experts the type faces found in a document. The hierarchical structure of the system makes it easy to integrate other experts as technology changes.

Recently, dictionary experts have been developed that allow omnifont systems to integrate context experts into the system. These and other lexical tools are discussed further in the Additional Features section in this chapter.

Omnifont systems offer moderate speed, which increases over time, and the greatest degree of accuracy and flexibility. The hierarchical structure of the experts and the self-learning process make omnifont insensitive to distorted text. OmniPage from Caere was one of the first products to use this technology.

ICR. Perhaps one of most easily confused terms in OCR vernacular is *intelligent character recognition* (ICR). Although used in different ways by several different vendors, ICR is commonly accepted as the general term for OCR using lexical analysis. Kurzweil, bought by XIS in 1980, was a pioneer in ICR and was probably the first developer to coin the term. XIS still uses many of Kurzweil's technologies and retains the name ICR for its proprietary OCR.

More recently, ICR has been extended to include handwriting recognition, often called *handwriting-ICR.*

Handwriting-ICR. Though available for 20 years, handwriting-ICR has only recently begun to develop into a commercially viable solution. "Constrained" handwriting recognition has been available for many years, but it recognized only characters drawn within a seven-bar box, very similar to those used in calculators and other LCD displays. Recently, "unconstrained" handwriting-ICR has been developed, which allows the recognition of relatively unconstrained characters (i.e., uppercase, noncontiguous, and confined to a field or a single line). Newer OCR systems have been released recently which can read lowercase and connected characters.

Handwriting-ICR should not be confused with stylus-based handwriting recognition, which is used with pen-based computers. A stylus-based system is significantly less sensitive and requires that each character be individually learned. A stylus-based system also relies on the strokes, speed, and sequence of the handwriting as it is being created. For this reason it cannot typically be used after the handwriting session. Traditional OCR can be used long after the information is imaged since it relies only on the image information.

Machine-Learned Fragment Analysis. *Machine-learned fragment analysis* (MLFA) is a technology developed by ExperVision, originally for the recognition of Kanji characters. MLFA analyzes each character as a set of fragments rather than as a single unit. The result is a much more detailed and therefore more accurate analysis of the characters. Currently, only a handful of companies provide products with OCR functionality that can be used for Kanji or Cyrillic characters.

Additional Features

Lexical Tools. OCR accuracy is enhanced with the integration of lexical tools that allow intelligent decisions to be made regarding characters that are not immediately recognizable. Possible translations for incomplete words are searched for in a dictionary or site lexicon. For example, if a term is perceived as being either "horn" or "hom," a dictionary would provide the insight necessary to determine that the word could not be "hom." If the use of a dictionary or site lexicon cannot provide a conclusive answer, grammatical and context-specific lexicon algorithms can be used. For example, if there is uncertainty over whether a term is "clog" or "dog," an initial check using a dictionary would not eliminate either possibility since both terms are valid words. By investigating the context of the word in the document a decision could be made in much the same way that a human operator might decide.

Trainable OCR. Trainable OCR works with both font-matching or pattern recognition as an underlying methodology to OCR. Trainable OCR provides a mechanism for the system to learn new fonts and characters dynamically. However, unlike omnifont, trainable OCR requires operator intervention. The system stores either the template or pattern rules for each character that is not recognized and prompts an operator for the appropriate character translation. Over time, the system becomes "smarter," and requires less operator input. The training process is often slow and tedious. Generally, trainable OCR is an efficient alternative to rekeying only when the type being "taught" will frequently reoccur in the text. Otherwise, the time lost in training the OCR applications will probably outweigh any potential time and cost savings.

Optical Mark Recognition. *Optical mark recognition* (OMR), also called *mark sense,* is one of the oldest and simplest of OCR technologies, often using the familiar scantron bubble-sheet. Requiring only a standardized form and a low-resolution scanner, the older applications of OMR continue to be very economical. For the same reasons, however, they are very limited. These applications use OMR to include check-box recognition into questionnaires and surveys. Unlike the traditional scantron, forms created with these products can be in virtually any arrangement, filled out in pen, and do not require bubbles or boxes to be completely filled in. The disadvantages of these forms over scantrons are slower recognition speed and a lower accuracy rate. Another application of OMR is the use of special *hieroglyphics* which provide various instructions to a receiving computer. These are used with "smart form" software, which is discussed later.

Double-Sided Scanning. The ability to scan and collate text from double-sided printed documents can save much time in situations where this form of text input is prevalent. A single side of each page in a stack is initially scanned and processed. The page stack is then flipped and the text on the other side is scanned and processed. The software automatically collates the text into its proper sequence.

Support for Columnar Input. Often, printed text is formatted in a columnar fashion (newspapers, for example). Typically, OCR "reads" text from left to right, top to bottom. Without support for columnar formatting, input of text in columnar format produces erroneous results where the electronic text, though technically correct, is stored in a semirandom order, all but destroying readability and comprehension. Scanner/OCR systems that support columnar text will recognize that the text should be processed left to right, within a column, top to bottom, before moving across the page to the right, to the next column of text. In this way, the text is stored in the order in which it was intended to be read. If your documents have columnar text but your scanner/OCR system does not recognize and automatically support this form of input, you must mask columns on the page and scan the document one time for each column, or OCR the document with one pass and then manually go back and reconfigure the blocks of text into their proper order.

Automatic Page Decomposition. Automatic page decomposition allows the OCR software to intelligently handle compound documents, documents that comprise images and graphics as well as text. Support for this feature is available in various degrees. Some products require templates to be manually drawn that separate and identify text and non-text sections of the document. The ability to automatically differentiate between images and text with a single pass of the document is supported by a large number of advanced OCR products. Simpler systems require that the document be processed by the scanner twice: once as an image and once with OCR software.

Automatic Conversion into Proprietary Formats. Originally, OCR output text as ASCII characters. Today, most products will store text directly into a specific proprietary format. This saves the user the time and effort of importing the ASCII text into another package (i.e., word processor or database). Additionally, the characteristics of the original text such as bold typeface, indenting, and underlining are preserved. Some products also support MS-Windows' *Dynamic Data Exchange* (DDE) which allows text to be stored directly into specific applications. Similarly, several "smart form" products allow data to be saved into databases, requiring little or no user interaction.

Neural Network Technology. By using fuzzy algorithms which help the software learn characters as they are recognized, "neural network" applications can handle a wide variety of typefaces and sizes. Beyond standard omnifont technology, neural networks are said to model the patterns of human thought by creating links between ideas and concepts. The primary advantage of neural network-based OCR is that it can recognize a wide array of documents, including those of very poor quality. The complexity of neural network technology requires a large number of computations; thus applications are relatively slow. However, when recognizing very poor quality documents, including faxes, neural network OCR is faster than standard OCR, and can recognize images that other products cannot.

TWAIN. *TWAIN* is an API standard that is intended to facilitate the integration of peripheral devices, such as scanners and optical drives, into an imaging system. Combined with a supportive scanner application, TWAIN allows scanning to be launched directly from a program such as Aldus' PageMaker. These products are often targeted to the novice or entry-level OCR end user. They completely integrate with text-based programs, including MS-Windows applications. This allows scanning and OCR to be performed directly from within a program.

Smart Forms. *Smart forms* is a new genre of product which uses ICR and OMR. These products allow remote data entry and PC control. Typically, a form is faxed to a computer with a faxboard and the requisite software, which responds to instructions based on a set of *hieroglyphics,* shapes which can be either partially or completely filled in to provide various commands. Many products provide platform compatibility with RDBMS by allowing information to be faxed from any platform, including handwritten text in many cases. By integrating fax machines as a primary input device, smart forms eliminate the need to rekey a large percentage of data.

Optical Disc Technology

One of the moving forces behind the popularity of imaging has been optical storage. (See Fig. 4-9.) Optical storage is not, however, a technology limited to imaging. Nor is there anything inherent in the image data type that requires optical storage. Optical storage can be a cost-effective alternative for any type of data. It is important to consider the speed, capacity, convenience, and cost of the full variety of storage devices before deciding on a specific technology.

There is certainly a natural fit of optical storage to imaging applications due to the large file sizes typical in imaging applications and the cost-effective storage capacities of optical storage devices.[13] Traditional magnetic storage can be far more expensive than optical storage in applications where instantaneous access to all images is not required. The liability of optical storage is that it has not caught up with the speed of magnetic storage. That creates a problem for applications that require the fastest possible retrieval speeds. But, as we will discuss, the decision to use optical storage involves more than just the issues of speed and capacity.

Optical Storage Alternatives. This discussion of optical storage will not delve into the technical details but will instead focus on the issues that you need to be aware of to evaluate and invest in this very volatile component of electronic document management.

There are four basic forms of optical storage: CD-ROM, WORM, erasable, and multifunction. Each form of optical storage has a specific purpose. Although some displacement has occurred in each category by developments in others, it is unlikely that any one form will be completely eliminated by the other alternatives.

[13]For an in-depth analysis of storage technologies and capacities, the reader is encouraged to read *Media Mania: The Fundamentals and Future of Removable Mass Storage Media,* Linda S. Kempster, Avedon Assoc. Inc., Potomac, Md., 1994.

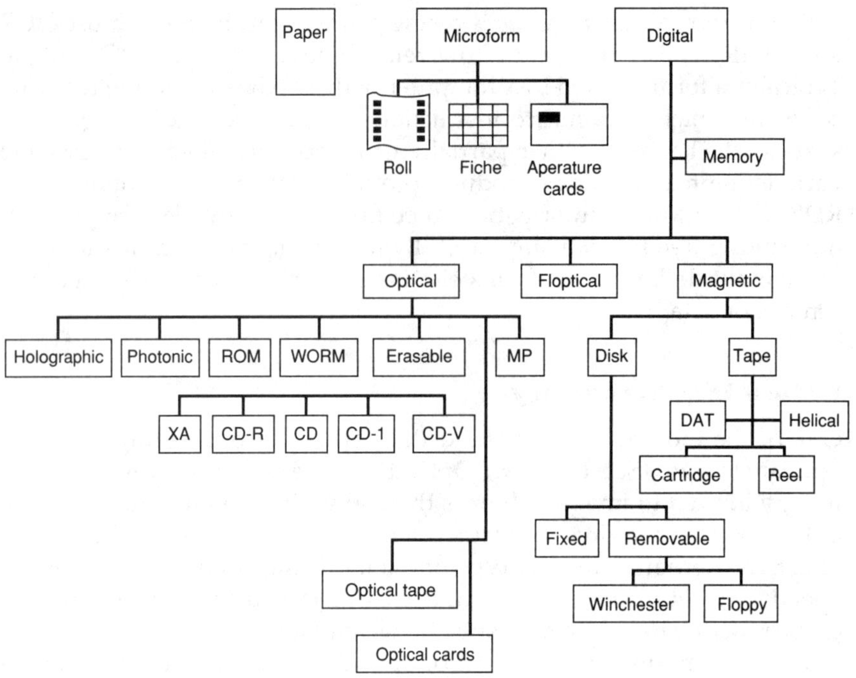

Figure 4-9 Optical document storage. Optical storage technology should be viewed as one component of a hierarchical storage architecture, where each storage media is used in a different aspect of the document management application.

Types of Optical Storage

- CD-ROM
- WORM
- Erasable
- Multifunction

CD-ROM. A publishing technology because of its inherent publishing quality, CD-ROM has been utilized primarily as a means to distribute large amounts of information that would typically be in hard-copy format, from periodicals, to research reports, to software manuals. CD-ROMs can be produced either through a third-party production house, which will press large numbers of CD-ROMs, or through the use of a CD-R drive, which allows individuals or organizations to simulate the pressing of CD-ROM by using optical technology that alters a CD-ROM–size disc without requiring actual pressing. Both approaches produce a standard CD-ROM that can be played on any industry standard CD-ROM player. The standard behind this interoperability is ISO 9660, also known as the *High Sierra* standard, perhaps the only true interoperability standard in the optical storage industry. Once a CD-ROM is created the publisher has the security of knowing that anyone with a CD-ROM drive can use and access their information.[14] It is the foundation and security that this standard provides that has led to the widespread acceptance of CD-ROM as a distribution media commercially. The CD-ROM sector of the optical storage industry accounts for approximately 50 percent of the entire optical market.

Even in this case, however, the "standard" can take on several revisions. Multimedia versions of CD-ROM, such as *Compact Disc-Interactive* (CD-I) and *Compact Disc Read-Only Memory eXtended Architecture* (CD-ROM XA), are supersets of the ISO 9660 standard which are not 100 percent compatible. Although the recordable version of CD-ROM, CD-R, can be played on any standard CD-ROM drive, it too is not 100 percent compatible. CD-R technology allows for multiple write sessions to the disc. Data cannot be erased from the disc once written, but unused portions of the platter can be written to in successive sessions.[15] This is unlike CD-ROM technology which writes to the disc once. Any portion of the platter not used is left "empty." Under the current technology, only that data written in the first write session is readable by a CD-ROM player.

Standards aside, one of the most crucial elements of CD-ROM production to consider is that of quality assurance and revision frequency. Once pressed, a CD-ROM cannot be altered. The last thing you want to do after pressing 40,000 CD-ROMs is discover a bug in the software, a problem with a hypertext link, or a faulty index. One shareware publisher we heard of discovered the existence of a virus in their CD-ROM after pressing several hundred thousand. The cost of repressing was prohibitive. As a result the CD-ROMs went out as they were with a notification that a known virus existed on the disc. You can imagine the reaction of users of the CD-ROM when they read that notice. What could the publisher have done? Certainly increase their quality control

[14]As long as the CD-ROM combines software for the target computer platform (Windows, DOS, Macintosh, UNIX) and the user has the appropriate drivers for their CD-ROM drive.

[15]At the time of this writing CD-R standards are in a state of flux, which has resulted in two types of CD-R Single Session, which allows only one contiguous write session, and multisession, which allows multiple write sessions.

efforts by testing the "publication" using an alternative updatable media such as CD-R or magnetic disk.

Remember also that the write-once feature of CD-ROM does not allow for updates and revisions without redistribution of the media. If you have any ideas along the lines of incremental updates that could be loaded onto a hard disk to supplement the original CD-ROM, think again. Users are not interested in the apparent convenience of such approaches. Studies show that a complete update that replaces the previous disc is by far a preferred alternative.[16]

Producing CD-ROM also requires the use of a retrieval application such as full-text retrieval or a database index. Some publishers create their own applications for this purpose but the trend is to use existing software-based retrieval mechanisms. One of the bigger problems in the CD-ROM market today is the lack of a standard search tool or front end to the CD-ROM data. Without such a standard, the user is compelled to learn as many different retrieval tools and languages as he has different CD-ROMs. Several attempts to develop such a standard are under way, but none has yet been widely accepted.[17]

Once the information has been created, organized, and checked for quality and the retrieval software has been determined and constructed, mastering can begin. The mastering process typically involves premastering an in-house version of the CD-ROM data and then sending a master tape out to an organization that will press the final CD-ROMs. The software used to create the premaster is often provided as part of a premastering package from the CD-ROM pressing service. The software ensures that all the files and directories on the CD-ROM have used ISO 9660 conventions. From that point on, the process is fairly mechanistic. A glass master platter is created in a clean-room environment. Then it is simply a matter of replicating the master.

An obvious question is why you would use CD-ROM when CD-R is now available. The answer involves cost and volume. The cost of the physical media and hardware necessary in each approach creates a threshold by which the CD-ROM pressed approach is more cost-effective for wide-scale distributions. A general rule of thumb to consider is that if you are planning on distributing more than 50 copies of each publication then CD-ROM technology is more cost-effective than CD-R.

WORMS. *Write once read many* (WORM) optical storage technology is similar to CD-R technology. It allows multiple write sessions, but once written to, that sector of the platter cannot be erased or reused. However,

[16]Delphi survey of on-line users of electronic information, 1992.

[17]Among the proposed standards for CD-ROM interoperability are CD-RDx and SFQL. For more information on these proposed standards see the section on query language in Chap. 5.

at 50 to 100 times the cost of CD-ROM, WORM is not intended to be used as a publishing technology. Its purpose was originally that of a cost-effective alternative to in-house magnetic media, but since the inception of erasable optical technology its appeal has eroded to a niche technology with little if any growth potential. Its primary use today is in the area of archival applications and the storage of legally sensitive information, which by virtue of WORM's noneraseability is best secured on WORM media. WORM technology uses a variety of technologies, each of which causes a permanent defect to be created in the surface of the disc. The defect may be in the form of a pit, a bubble, an alloy, or a change in the actual media which cannot be reversed. To retrieve the information, a low-intensity laser is used. The light reflected from the disc's surface is measured—a defect will cause the light to be refracted and a portion of the disc that does not contain a defect will reflect the light. The defect is interpreted as a binary 1 or 0, depending on the manufacturer. (See Fig. 4-10.)

Because of the time required to perform the physical pitting of the disc and the bulk of the laser optic assembly, WORMs are considerably slower than high-performance magnetic disks. One of the side effects of the permanent pitting is that old versions of files can be retrieved indefinitely if the application provides for version control.

There are two additional forms of WORM technology. Although these approaches differ radically from traditional WORMs in the physical form of the media used, they use similar read/write technology. The first approach, known as *optical cards,* repackages WORM into a popular consumer metaphor, that of the credit card. Optical cards, developed by Drexler Technology and Canon, have the ability to store several megabytes of data on a reflective card no larger than a standard credit card.[18] This technology is being used commercially but it has been slow to catch on. Part of the problem is the concern over a lack of standards and changing form factors.

The second alternative form of WORM technology is *optical tape.* Developed by ICI, optical tape comes in several form factors including disc, card, tape cartridge, and reel tape. The 12-inch reel is almost 1 kilometer long when unwound. It can hold 1 terabyte of data and can be passed through a tape drive in less than 60 seconds. Optical tape provides an ideal means of backing up optical media for off-site storage. A single tape can hold more than fifteen hundred 5.25-inch erasable optical discs, which allows the discs to be reused while the archive information is unchangeable since it is on WORM media.

[18]This should not be confused with *smart cards,* which incorporate actual microprocessor technology. The optical card is strictly a storage medium.

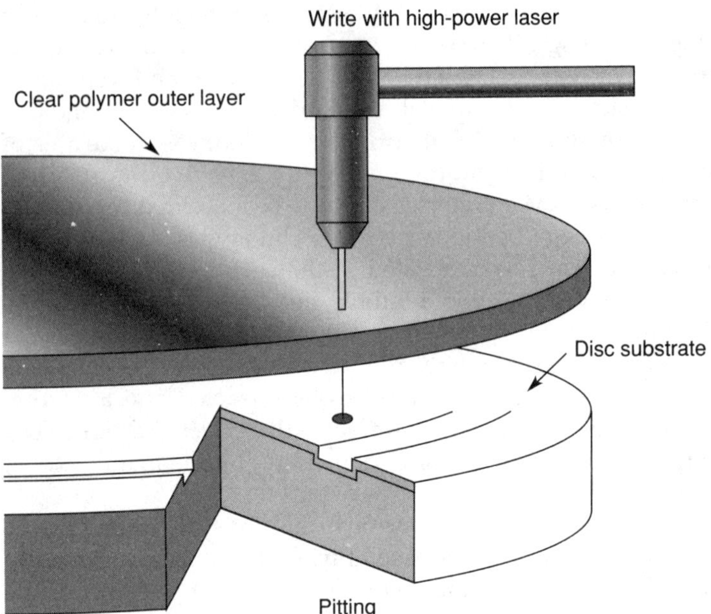

Figure 4-10 WORM mechanics. WORMS are written to by using laser optics to create defects in portions of the disc's surface, called *domains.* To retrieve information, a lower-intensity laser is used. The light reflected from the disc is measured—a defect in the platter would refract light and thus register a lower reading. The process shown is called the *ablative method.* It requires an air gap above the pitted material so that a pit can form. Because of the air gap, corrosion may occur over time causing degeneration of the write surface. Sony uses an alloy method that heats layers of separate metals and causes them to form an alloy in the heated spot. This affects the material's reflective properties in such a way that the laser can read the change in the bimetal and alloy surfaces. (This figure represents only one side of the optical disc. Both sides would be identical in composition and capable of recording data. However, the disc would have to be physically removed from the drive and turned over. This is done either through a robotic or manual operation.)

Erasable Optical Disc. *Erasable optical discs* work in very much the same way as WORM, but allow for update and erasure of data. As with WORM, there are a variety of alternative technologies used to read from and write to erasable optical media. Each provides for a method of changing the state of data on the disc without causing a permanent defect to the disc's surface. The magneto-optical technology is shown in Fig. 4-11.

Multifunction Optical Storage. The latest development in optical storage technology is that of multifunction media. *Multifunction optical storage* combines the benefits of WORM's security and the flexibility of

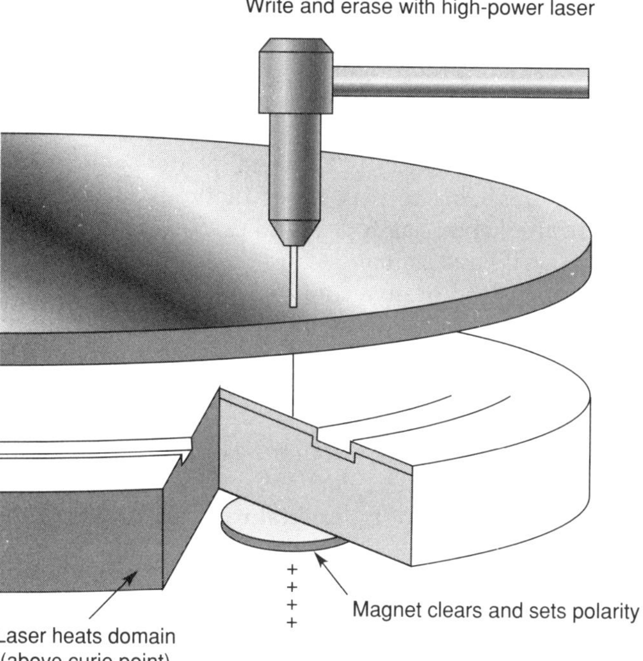

Figure 4-11 Magneto-optical mechanics. This approach to erasable optical disc technology is the most popular available. In this approach, a laser is used that heats the disc's media above its Curie point. At that moment, the disc will take on the polarity of the magnet that it is sitting on. By reheating the disk and changing the polarity of the magnet, the disc can be reset any number of times; thus, it is "erasable."

erasable technology. There are two distinct types of multifunction optical storage. The first allows both WORM and erasable discs to be used in a single drive. This minimizes the expense in hardware but does not eliminate the need for two types of media. The second approach uses a single drive and a single media type. The media is factory encoded in such a way that part of the disc is write-only and part is fully erasable. This introduces yet another standard into the already convoluted standards arena, but it is a cost-effective means of preserving the most attractive features of each media type.

Optical Juke Boxes. *Optical juke boxes* provide massive storage capacity by housing large numbers of optical platters that can be played on one or several spindles (turntables) when information on the disc is requested by a user. Here, too, a variety of mechanisms and media formats are supported. The media used in an optical jukebox are often ruggedized as well, in order to survive the constant abuse of being swapped in and out of their housing. Nonetheless these are very reliable devices for mass storage.

The problem to be considered thoroughly when investing in an optical jukebox is that of intelligent configuration and access. Juke boxes require intelligence. Think about it this way: A juke box contains a certain number of optical platters, let's say 100. Optical platters will always outnumber the spindles available to play those platters. For example let's say our hypothetical jukebox has four spindles. (If that wasn't the case, then you might as well have as many stand-alone optical drives as you have optical platters, instead of investing in a jukebox.) That means that only four discs can be used at any one time. If there are 100 users attempting to access the juke box on a purely random basis, 96 of them will be put into a queue and their wait time can become significant. In cases of poorly configured juke boxes the queue time could stretch to minutes. That will depend on a variety of other factors as well, such as the number of pickers (the robotic arms that actually retrieve and load the disc),[19] access time, distribution of individual requests, and the time required to actually keep the optical disc on the spindle. But the most important of all factors will be the intelligence used to configure the information stored on each disc and the algorithms used to perform functions such as caching of information rather than keeping a platter active, or using a look-ahead algorithm to read additional information that has been defined as a probable request in further processing, rather that reloading the same disc back to back. The specifics of these algorithms can be many; what is important to remember is that some degree of intelligence is required in order to adequately implement the juke box component of the solution.

Standards. Each approach to optical storage has its own set of standards. Even within each technology there are conflicting standards as to tracking methods, writing conventions, and media size. With all of these options available, the issue of standards plays a significant role in the optical storage industry. There are more than 20 separate standards for optical storage technology.[20] None of these standards offers interoperability. The problem this creates for many evaluators and implementors of optical storage is not only deciding on a standard approach internally but also determining how the variety of imaging systems used in an organization will work together. Much of the attention paid to optical storage technology focuses on this issue of standards. Table 4-1 provides a summary of current optical standards at the time of this writing.

[19]In some newer juke boxes, the read–write head actually moves instead of the platters, but there are still a limited number of read–write heads.

[20]Fifteen of these storage standards are shown in Table 4-1.

Table 4-1. Optical Standards: A Maze of Options

Form factor	Capacity	ISO	ANSI
14-inch WORM (Kodak *only*)	3400 MB/side	10885	X3.200-1992
12-inch WORM	(No standards)		
8-inch WORM	(Used in Japan *only*)		
5.25-inch WORM	325 MB/side	9171A (CCS)	X3.211-1992
(Pioneer *only*)		9171B (SS)	X3.214-1992
	650 MB/side	NA	X3.191-1991 (SSZ)
5.25-inch MO (Same as ISO 9171)	325 MB/side	10089A (CCS)	X3.212-1992
		10089B (SS)	
5.25-inch MF	325 MB/side	11560	X3.220-1992
(Same as ISO 10089A but partially rewritable)			
5.25-inch MO	325 MB/side	(ECMA 183 Standard *only*)	
5.25-inch MO	650 MB/side	(ECMA 184 2X Standard *only*)	
5.25	975 MB/side	SC23 (No manufacturers yet)	
5.25	1.3 GB/side	(Sony, no standard)	
3.5-inch MO	128 MB one side	10090 (CCS)	
3.5-inch MO	120 MB one side	NA	X3.213-1992 (CCS)
(No manufacturers yet)			(Discrete Block)
3.5-inch MO	30 MB one side	(IBM, no standard	

It should be pointed out, however, that media standards have always been in a state of flux. Even in the recent past, diskette sizes and capacities have changed in ways that create issues of incompatibility. In fact, the only standard that has withstood the test of time in the computer industry is that of the 800–1600 BPI magnetic tape. Since it has been used as an archival standard, the magnetic tape has been regarded as a long-term storage media. As a result, technology demands for its form or functionality have not changed from year to year. Optical storage serves a vastly different role. It is used for a variety of applications—from on-line to archival. As a result, the demands for a broad range of functionality and uses have allowed vendors to target niche markets for the technology, creating a vast array of form factors, speeds, media formats, capacities, and resulting standards.

Until the standards for optical media evolve further this will be a persistent problem. Evaluators have little option but to balance the potential obsolescence cycle of optical media with the payback offered by an imaging system, since no security exists to ensure that any single standard will prevail for long-term storage. The irony is that many of the optical storage devices on the market today claim shelf lives[21] of

[21]Shelf life of optical storage is up to 600 years for CD-ROM.

up to 100 years. The rate of technology change virtually assures us that most of these same vendors will not be here to support their technology even a few decades from now, much less in 100 years. It is absurd to expect that any technology can be viable for such a long period of time. That does not ease the burden on many users, especially those with legal or regulatory concerns. Obviously this creates an archival and records management dilemma due to the importance of ongoing access requirements. The only solution is to consider the available options and the potential longevity of the respective vendors along with the specific requirements for long-term storage. We can also anticipate that new technologies and methods will come along to displace optical storage in the same way that optical threatens to displace magnetic.

Even as you consider the impact of existing storage standards on your imaging systems, developments are under way that will shatter any of these current approaches. One example is that of *photonic storage,* a technology developed and patented by a team at the University of California, Irvine, headed by Dr. Peter Renzipis. Photonics has the potential to completely alter the face of storage within the next decade. Unlike all traditional storage media to date, these devices do not require any moving parts, making retrieval time virtually instantaneous. The storage capacity of a crystal smaller than the size of a sugar cube is 2 terabytes. A similar technology under development by IBM provides for 1 terabyte of storage on a space the size of a device that could pass through the eye of a needle.

Within the next five years these technologies will be available on integrated circuitry. The issue of cost-effective storage at that time will be a moot discussion. But what are the implications of having such phenomenal storage capacities available? Does it dissipate the problems of electronic document management? Ask that question of users who are already overburdened with information in electronic repositories that are impossible to access. Ask any records manager trying to develop a plan to define a life-cycle for all of the organization's documents. You must carefully consider and design retrieval and management systems. One of the more powerful of these approaches, Text Retrieval, is the topic of the next chapter. The problem of managing electronic documents is far more complex than just providing storage. It would be foolish and naive to think otherwise.

5
Text Retrieval

What Is Text Retrieval?

The advent of text retrieval technology over 20 years ago represented a significant difference in on-line document management. For the first time, a method of retrieval based on the full document content was made available. Today, text retrieval is the most widely used form of content-based retrieval document management. The predecessors to text retrieval, keyword systems and document abstracting, start with the full content of the document, but distill the document's content down to a particular list of words or rewritten paragraphs, respectively. Thus the researcher's view of the value of the document is immediately prejudiced or limited by the content-based methodology. This is not the case in a text retrieval system.

In its simplest form, text retrieval is a document management system that can reference and retrieve structured or unstructured textual information such as memos, manuals, contracts, letters, or news stories, based on words and word meanings found within the full content of the document. (See Fig. 5-1.) Unlike databases that support variable-length character fields, text retrieval systems have no practical limit to the amount of text that can be stored within a document. In this way, text enables the single-point-of-access environment by providing a means to locate and retrieve virtually any and all information resources pertaining to a particular area of interest, without the need for a predetermined access strategy and the distillation of information innate to these approaches.

But text retrieval is not a panacea. It is a toolset that enables developers to provide an environment to exercise tighter and more powerful control over the precision and recall effectiveness of the research-retrieval process. Therefore, in order to properly assess your need for a text retrieval system, and to adequately compare product approaches, it is critical that an understanding of the difference between precision

"In order for knowledge to be used it has to be selected out of an undifferentiated mass. Knowledge that cannot be selected is knowledge lost in vain".

Figure 5-1. Illustrated quote from Vannevar Bush (1945).

and recall, and the type of research that is to be conducted in the resulting environment, be understood. (See Fig. 5-2.) *Precision* is the ability of the system to provide *only* that information that is relevant. On the other hand, *recall* is the ability to return *all* the information that is relevant. But these two measurements are not mutually exclusive. It is striking an acceptable balance between the two that is the task before the developer of a text retrieval system. Consider, for example, an on-line collection of 500 documents. A system with 100 percent recall capability is simple to achieve—no matter what the query, the system would return all 500 documents. Perfect recall is achieved, all the relevant documents are retrieved—and a whole lot more (precision was not considered).

This concept of precision versus recall is innate to any structured approach to retrieving unstructured information. Thus, this design problem is one that is not new to individuals such as librarians. But for those individuals who have become accustomed to traditional database approaches to data management this is a new issue. Traditional structured databases comprise a finite set of data fields each containing a finite set of values. Research results are controlled and predictable. For example, in a human resources traditional database, to locate all employees that work in the "sales" department, a query that extracts those records in which the "department" field equals "sales" would give perfect results (all those employees in this department, and only those employees). But consider the problem with searching an on-line collection of personnel documents (resumes, performance reviews,

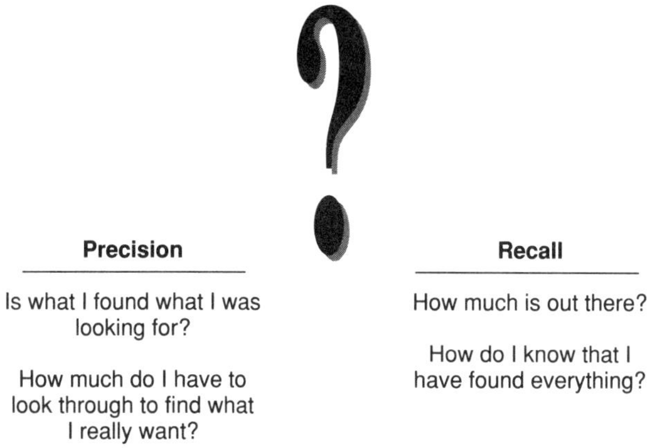

Figure 5-2. The information retrieval dilemma: precision versus recall. The effectiveness of a text retrieval system is measured by its ability to control precision and recall. These two properties of the system represent different directions that research may take based on the type of query they are executing. The objective of a well-designed system is to strike an acceptable balance between the two and to offer tools for increasing one or the other.

letters of recommendation or reprimanding, etc.), looking for those employees that exhibit potential for establishing a new customer support department in the Tokyo office. In this case there is no single field or finite value that you can retrieve on. Rather an evaluation of the contents of the documents must be made and "decisions" executed determining whether this employee exhibits such qualities. The process becomes much more subjective, and thus precision and recall become an issue.

There are various methods and tools for controlling a system's level of precision and recall, and these will be discussed in this chapter. But before the tools are introduced, it is important to appreciate that the needs for precision and recall control are dependent on various factors including the size of the document collection, users' familiarity with the contents of the documents, and the types of research that are conducted.

Text Retrieval Research Environments

There are three basic categories of research that can be supported through a text retrieval system: finite research, investigative research, and serendipitous research. Each defines a category of text retrieval functionality and control required. Each one also represents an increasing level of complexity in the text retrieval process.

Finite Research

This is the simplest and most easily implemented form of text retrieval. In this environment, the users possess a high degree of knowledge about the contents of their documents. They bring to the research process a clear and concise definition of what they are looking for, and knowledge that it exists. This environment is known as finite because once the suspect document is found the research is done. To draw an analogy with the needle in a haystack, finite research always has a given number of needles in each haystack—no more, no less. If you know there's one needle, the problem is only the time allotted for the search, because you can continue looking until you find that one needle. Once you find it, your job is over. Finite research is very much the same. Where text retrieval helps is in eliminating the research time.

For example, a user remembers filing a document in which the phrase "the criminal was apprehended red-handed" is contained. But the document is somewhere among several hundred others, and the user can not recall the name of the file or the date on which it was filed. A search for documents that contain this exact phrase will quickly provide the document that the user is looking for. The focus in this type of situation is in precision—an exact file is required. But this situation of 100 percent precision is achieved only when the user can succinctly and explicitly express what they are looking for in the query.

In other situations, although the user is looking for a finite set of documents, the request can not be expressed concisely or succinctly. Although these cases are a form of finite research, because a particular answer or set of documents is required, the environment is somewhat different and requires a higher level of intelligence from the text retrieval software because the user's knowledge is less than 100 percent. For example, a user is interested in finding all documents that pertain to cancer research. In order to provide necessary recall, the user is expecting the software to understand that the concept *cancer* can be manifested in text in a variety of ways (leukemia, malignant tumors, etc.). Although a fair degree of precision is still necessary (do not return all documents), the focus is on recall—find everything that pertains to this subject, not just the word or phrase itself that the query contains.

Investigative Research

A more complex research model is investigative research. In investigative research the research process becomes more interactive. The researcher has no knowledge of what is out there or the best way in which to express the request for information. In fact, the researcher may only have a vague idea of what he or she is looking for. Time cannot be compressed merely by speeding the search of the documents; it must be compressed by assisting the researcher to interrogate the document database in an iterative fashion. In this way the researcher will gain an understanding of the document collection's contents and be able to better determine what, if any, relevance may exist in its contents. For example, a user may request information on "missing data." The system would suggest that "missing data" is associated with several factors: viruses, illegal electronic file transfer, software piracy, etc. The user would identify "viruses" as a topic of interest. The system may then respond with a list of known viruses. The user could choose any or all of these as a subsequent avenue of research. This iterative process continues until the user is satisfied with the collection of documents presented.

Assistance from the software in defining a valid request and subsequently locating and retrieving the proper document collection is required. In these situations the focus is initially on recall, used to broaden the user's perspective and choices, but the interactive process assists the user in focusing on precision, eventually leading to a finite set of documents. Keep in mind that investigative research is not haphazard or undirected. The objective is clear to the user that a particular piece of information will satisfy their research. In that regard it is closed-ended and has a point of conclusion if for no other reason than the time allotted to the research task. This is in contrast to serendipitous research.

Serendipitous Research

In an environment of serendipitous research there is no specific concept, idea, or destination for the research. The researcher has no direct interest in either the amount of information available or the relevance of information to one specific concept. Instead it is the linkages of concepts and documents one to the other that the researcher will use to navigate through the document collection. This may sound a bit unlikely, but stop and consider the value of serendipitous research throughout the history of mankind. Many of the greatest inventions we live with on a daily basis are the result of such serendipity in research. Velcro, quinine, smallpox vaccine, polyethylene, Teflon, and a host of other discoveries can be attributed to the randomness of research.[1] Serendipity is very much part of the way we conduct research. What is interesting, but should also be surprising, is that text retrieval can facilitate and even encourage serendipity by allowing researchers to make more numerous connections among apparently unrelated events and information stored in document collections. If you consider that it may not be so far-fetched to imagine that text retrieval technology could be the gateway to an era of accelerated invention and discovery. Consider the pharmaceutical company whose marketing department uncovered information that had eluded the R&D department for years. This marketing group discovered that a drug to control hypertension, which was based on a chemical compound for which the company had a patent, also caused weight loss in humans. The marketing department, with a different slant on the value of such information than that of the R&D group, decided to repackage and remarket the compound under another product name targeted at a different market—dietary aids. In this way the company is getting twice the bang for their buck, by marketing the chemical compound to two distinct groups, each with their own market potential.

Although this data had existed for years, it was under the control of the R&D group, who do did not view the information in quite the same way as the marketing group. But, it wasn't until a text retrieval system, complete with a user-friendly intuitive front end, was made available that users other than R&D's scientists could begin to make heads or tails of the masses of scientific information amassed through R&D's efforts.

Opening doors to information for haphazard searching by virtually any type of inquirer holds untold promise for the value of a text retrieval system. Perhaps the answers to many of our most agonizing research endeavors, cures for diseases, and the answers to untold scientific questions can be unlocked through such power. It is not unthinkable; it just requires that the raw data be present and that the research tool provide a dynamic level of control of precision and recall allowing users to direct searches in certain areas, and then broaden and narrow further investigation as the case may warrant.

None of these categories of research and the requirements they place on a text retrieval system are meant to be exclusionary. Indeed, they are not. Undoubtedly the first two will exist in most all organizations. The third is predominantly associated with more intensive research

[1]*Serendipity: Accidental Discoveries in Science* by Roberts, Wiley, 1989.

environments such as laboratories and think tanks. But stop to consider the extent to which we are all stumbling across new information and making connections on the fly among documents and data we encounter in our day-to-day tasks. The value of a tool that facilitates this process is just beginning to be understood by most of us. Each research environment and the text retrieval tools required represent a necessary element of enterprisewide document management that needs to be supported by text retrieval.

A Slowly Evolving Technology

Considering that text retrieval can support each of the environments described in the previous section, it is interesting to note that the technology has been less than enthusiastically embraced by a large market. Text retrieval technology has been commercially available for over 20 years. The technology was originally developed by organizations and institutions for proprietary applications. The first of these available to the general public was STAIRS, developed by IBM. STAIRS was developed internally to manage extensive litigation support documentation.[2] Other pioneer products in the marketplace included Information Dimension's BASIS (originally developed by Battelle Laboratories), BRS Software's BRS/SEARCH, InfoData's InquireText, and Data Retrieval's Text DBMS. Despite the power of these systems and their potential to revolutionize the science of electronic document management, growth over the past two decades in this segment of the software industry has been especially slow.

One must appreciate that the initial product offerings supported only the finite research environment. Thus they required that the work environment be somewhat controlled and used by a highly knowledgeable group of researchers. The advent of additional enhancing features such as synonym tracking and heuristic association, which are discussed later in this book, made text retrieval systems more powerful and appealing. But, equally responsible for the slow adoption of this toolset was the lack of complementary technologies needed to make text retrieval a cost-effective solution.

It was not until the mid- to late 1980s that these complementary technologies, including electronic desktop publishing, word processing, optical character recognition (OCR), and imaging, became widely

[2]IBM originally developed STAIRS as a way to aid document retrieval in an antitrust lawsuit against Control Data Corp.

accessible. The increasing creation and storage of electronic documents as the result of the reduced costs and increased efficiency of desktop computing, optical scanning devices, and OCR, along with more efficient storage devices such as optical discs, made text retrieval a viable technology for the electronic document. With these technologies penetrating the marketplace, the growth prospects for text retrieval have accelerated substantially. Text retrieval has become not only an effective method but in most organizations a necessary component of an EDMS.

Today the application of text retrieval is becoming a significant competitive tool for organizations attempting to differentiate themselves. Applications of text retrieval abound. Few industries have not used text retrieval in some form. Most often the text retrieval system is a cornerstone component of an application, the reason being that text retrieval can drastically alter the ability to work effectively with large quantities of knowledge in document form. Without the text retrieval function this same information would be an unwieldy resource, incapable of supporting timely decision making. One of the best examples of such an application is Corning Glass Works' competitive intelligence application of text retrieval. Another strategic use of this technology is as a router of information. That takes many forms in many applications, but the most common aspect of its implementation is a subtle change in the organization from a distributorship model of information transfer to a demand-based, user-centric model, similar to what we had originally described as the point-of-research model of information systems. This model of receiver-driven information has the effect of collapsing business cycles and speeding the responsiveness of service organizations. An example of this approach to text retrieval implementation is Liberty Mutual's application of text retrieval in the insurance industry. Therefore, before we discuss the technology behind these types of applications, let's first take a look at these showcase examples of the types of results organizations are achieving through the deployment of text retrieval technology.

Case Study: Corning, Inc.

Corning, Inc., formerly Corning Glass Works, is a worldwide manufacturer with $3 billion in annual sales. It produces more than 60,000 products and has four major business segments:

- Specialty Materials produces emission control substrates, headlight lenses, and ophthalmic glass.
- Communications produces optical fiber and television display products.

- Laboratory Services conducts clinical, laboratory, and environmental testing.
- Consumer Products produces cookware.

Corning, Inc. also has over 20 joint ventures which provided approximately 40 percent of net income in 1990.

Corning identified the lack of sharing of information among business managers as a major business problem. To attack the problem, corporate marketing formed a business information *corrective action team* (CAT), comprising representatives from most of the major business units in the company. The team began by conducting an information audit consisting of interviews with 70 key managers and executives. This audit not only identified some of the costs attributed to the lack of information sharing, it also served as the foundation for a thorough understanding of the nature of the problem and of the users' perspective on what types and forms of communication were needed.

Tangible evidence was documented in which the lack of sharing of information resulted in the loss of millions of dollars in sales to competitors. An incident was uncovered that subsequently played a major role in the justification of a system for information sharing. The impact of this incident was so large that the lost sales were roughly equivalent to the annual revenue of one operating division. One executive interviewed stated his belief that 70 to 90 percent of major business mistakes at Corning were due to a lack of competitive information. In the words of Corning's manager of information management in corporate marketing, "Senior management felt a solution to the information sharing problem was an essential cost of doing business, almost as basic as having security systems in all buildings."

User perspectives uncovered in the audit were instrumental in designing requirements for the system which was subsequently implemented. One of the primary findings was that the flow of general business and competitive information across department and division lines was infrequent and informal. Business managers sent information only to others whom they knew personally, and then only if they did not need the information themselves and if they knew the recipient had a need for the information. Managers at Corning felt a strong need for information that was easily accessible, accurate, and timely. Abstracts were considered unacceptable since authorship was questioned and readers' skepticism caused them to want to view deleted material as well.

Any new system had to be easy to use and flexible. Additionally, most managers insisted on having direct control over their portion of any database to be established. This was even more important if the database was to be centrally located. Managers also expected the program to supplement and assist rather than duplicate what they were already doing. All of this indicated the need to develop a means for exchanging information while maintaining local control. The ideal solution was a combination of inside and outside information that used the existing E-mail network.

Corning developed an objective to improve the communication of textual business information and took an innovative approach by shifting the bulk of the responsibility from senders to receivers. This was in line with the corporation's philosophy that competitive information is everyone's responsibility. Corporate marketing implemented a system that targeted marketing business managers as the primary users, but in concept, any user of the corporation's worldwide E-mail network is able to access the system.

Since the information audit had successfully articulated the business issues and identified the user requirements, the evaluation and purchase decision process associated with the new system took only weeks. The MIS department recommended a relational database solution as a result of their familiarity with it; however, that technology was rejected as insufficient when compared with user requirements. One of the biggest factors in the product selection process was the user requirement to capitalize on their familiarity and use of the existing E-mail system.

Corning designed the Business Information Exchange Network which runs on an enterprise network with text retrieval vendor Henco Software's Synchrony product.[3] It is a menu-driven, full-text database operating as an integrated component within Corning's existing E-mail and office information systems. The network has the following characteristics:

- Menu-driven for ease-of-use
- Tied into E-mail, shared folders, and scratch pad features
- Clipping folders that contain the results of nightly, customized searches

When users choose to make an entry to the network database, they do so through an already familiar E-mail interface. This permits easy access from anywhere on the network and sharing of internal documents without converting them.

[3]In 1993, Henco Software changed their name to Expressway Technologies; it no longer supports or sells the Synchrony product.

A key feature of the network is the ability to establish clipping folders set up by each user based on his or her interest profile. The database automatically performs searches nightly using these personalized information requests. The next day a message informs the user of the result. This amounts to a customized electronic magazine.

Users can also review from the main menu all information added within the last five days, which is like reading a newspaper since the information is not selected solely based upon the reader's interest. Additionally, users can enter English language commands to conduct an ad hoc search of the database using Synchrony's boolean logic.

An analyst can develop an outline for a report and then conduct a search based upon each line of the outline. As each search result appears, the analyst can save selected information on an electronic scratch pad for later editing. Since the saved screen also contains title and source information, it makes reference citing simple.

Corning claims that over 50 percent of the information needed for decision making already exists internally in such items as memos, trip reports, and "unconfirmed information." Pride of ownership is an important factor in motivating users to contribute entries, so originators' identities are retained. External information such as news articles, trade journals, and other publications are generally input electronically from databases and electronic news services.

The implementation process was heavily user-oriented, with considerable customizing and piloting. Currently 12 business units are in operational mode with over 20 "topics" implemented as well. *Topics* are databases requested by users covering particular areas of interest which may overlap business units, such as the recently opened database on China. There are a total of over 30 databases, with further expansion planned. The number of users of each database ranges from a minimum of 8 in certain topic areas to as many as 160 in a division.

The success of Corning's system is best reflected in its impact upon the way users work. Users receive highly targeted information on a daily basis and generally need not spend time searching, thus improving productivity dramatically. As an example, business managers who recently changed jobs reported that usage of the system allowed them to get up to speed in hours rather than days, as it would have previously.

When users feel secure with the information they've read and feel they haven't missed anything, they are empowered with immeasurable confidence which results in improved responsiveness. Therefore, when a question is presented, a high-quality decision can be made more quickly and easily without time-consuming and costly research.

Furthermore, Corning is leveraging the formerly untapped resource of information held by individual employees. The system fosters increased motivation to contribute information by freeing the contributor from concern about whom to address. This is important in today's demanding business environment in which corporations must have high levels of internal information sharing in order to be competitive. Corning's philosophy is that competitive information is each employee's responsibility, and the corporation has succeeded in implementing a system which empowers each employee to rise to that vision.

This theme of empowerment is one that you will often see expressed in text retrieval applications. It is especially evident if you look at the enthusiasm of text retrieval users. Despite the longevity of the text retrieval industry, those working with it for the first time are amazed that it has not become a mainstream technology sooner. That slow rise to prominence, outside of the library sciences community, may be the most frustrating and often-discussed phenomenon of the text retrieval industry.

Case Study: Liberty Mutual Insurance

Based on consolidated assets of $29.1 billion, Liberty Mutual Insurance Company is one of the nation's leading providers of personal, business, and life insurance; international services and products; and financial services. It employs more than 19,000 people in over 400 locations throughout the United States and Canada. In 1992, Liberty Mutual celebrated its eightieth anniversary.

In 1989, Liberty Mutual began implementation of a significant, new strategic plan that would direct its long-term use of information technology. The backbone of the plan became a multiplatform client/server computing environment employing a mix of some 10,000 intelligent workstations (most of which are Apple Macintosh computers), a host of midrange processors and several mainframes, and an integrated data network extending to some 250 sites and linking computing resources with more than 15,000 users. The company also maintains its own voice and telecommunications networks. It is important to note that Liberty Mutual is strongly committed to this client/server environment and has established a policy that software for its user network be Macintosh compatible.

Senior management chartered the information systems group, located in Portsmouth, New Hampshire, with fueling the advance of technology in the corporation. In support of this objective, the information systems group established and funded its own research and development group. Unlike the rest of information systems' workgroups, which deal with day-to-day issues impacting Liberty Mutual's business, the R&D group identifies longer-term needs and solutions with emphasis on flexibility in sharing information, increasing the cost-efficiency of the organization and its systems, and encouraging excellence in customer service. To date, R&D has pioneered advanced workflow systems, corporate information management systems, sales workstations, and customer risk information systems for Liberty Mutual.

In mid-1990, an enterprise-wide information needs assessment conducted by corporate administration found that redundant requests for external information and inconsistent access to dated internal documentation was a recurring problem. For example, claims agents across the country routinely cross-referenced internal policies and procedures; state and federal OSHA and insurance regulations; and financial information from A. M. Best, Moody's, and Standard & Poor's. Much of the information was in hard-copy form. Managers and customer service personnel strained to keep up with the volume and variety of detail relating to changing health care regulations and competitive market developments.

For the information systems group, the results of the needs assessment confirmed its concerns about the quality of information and the way people used information delivery systems. The results of the assessment came at a time when the users were becoming more proficient in working with IS tools available at their workstations.

Various departments asked for better ways to find specific information. IS felt the availability of text retrieval over the network had the potential to stimulate better dissemination of information and reduce the cost of acquiring and managing information. In the long term, IS believed text retrieval would bring significant competitive advantages and help shorten business cycles.

R&D's recommendation was to offer text retrieval tools in a receiver-driven system. With such a system, an individual could request information that is filtered to personal specifications and have the information dynamically delivered to his or her workstation. This approach would reduce or eliminate the need for paper documents and the problems associated with distributing updates that keep paper documents current. The system would also provide a consistent, single point of access to both internally and externally produced textual information.

R&D proposed a system they called an *information refinery.* Information would arrive from multiple sources, internal and external, and subscribers could extract the relevant information they needed based on their own criteria. The refinery would provide the framework for managing textual information and integrate text retrieval into the business workflow. In addition, to minimize costs and potential learning curves, the refinery sought to use existing technical components, such as network and E-mail services, where possible.

In developing the information refinery, R&D looked not only at the readiness of the work force but also at the sophistication of products available in the industry. Several years earlier, Liberty Mutual had evaluated text retrieval for its network and found the products more appropriate for special-purpose applications rather than general use.

IS did develop several text applications; however, it did not find a general-use text retrieval product compatible with Liberty Mutual's client/server architecture. That changed in 1990 when Verity, Inc., the developer of the TOPIC text retrieval product, began development of a Macintosh client for TOPIC and announced TOPIC REALTIME for DowVision (which interfaced TOPIC with the Dow Jones News Service).

Believing that the real benefit of text retrieval would come from its general use network-wide, R&D saw TOPIC REALTIME as the way to demonstrate the potential of text retrieval in delivering information, as it happens, to the entire organization. Typically, text retrieval enters an enterprise as a pilot application in a small, well-prepared user community. Liberty Mutual made its information refinery available to the entire user network—from technology pioneer to casual user, from senior manager to field agent. Because widespread organizational acceptance was a goal, R&D specifically designed the initial release of the information refinery for a passive user—an individual who wants timely access to predetermined types of information yet has no knowledge of text retrieval products and wants to expend little or no effort to get the information.

The information refinery captures documents from DowVision, a group of 24-hour news wires that Liberty Mutual receives via a feed over a dedicated X.25 line. Dow Vision includes the *Wall Street Journal,* Dow Jones News Services ("The Broadtape"), Business News Wire and PR News Wire, Capital Markets Report, Dow Jones International News Services, Professional Investor Report, and Federal Filings. The first releases of two text retrieval services, called InfoWatches, were rolled out in January 1992. These services are being readied for Liberty Mutual's internal E-mail systems. Eventually, each of Liberty Mutual's 10,000 E-mail users may have information at their desk within minutes after the news is off the wire.

The text retrieval system provides the filtering process. The topic-based approach caught the interest of IS because it gave the user the ability to set the appropriate degree of interest in a topic. If casually interested, the user would set a high threshold and see only one or two stories a day. If very interested, the user would set a low threshold and see every story that referred to the subject.

Liberty Mutual's IS R&D staff used the system to write the queries that drive the two InfoWatch services, a managed care watch and a technology watch. They interviewed internal experts on the two watch subjects and built extensive lists of terms, vendor names, products, and key issues. Each term was assigned a point value based on its importance. These preassigned values are used to assess the relevance of stories arriving over the news wire.

The technology watch tracks about 150 terms dealing with technologies, vendors, and products. The managed care watch evaluates over 60 terms relating to health care costs, providers, strategies, and organizations. Of the time it took to write the two InfoWatch applications, approximately 20 percent was spent identifying and researching terms, and another 20 percent was spent on data entry. R&D spent the remainder of development time fine-tuning the TOPIC application, particularly in refining the relevancy thresholds so users would get accurate responses to their information requests.

Anyone on the network can use either InfoWatch by setting a personal interest threshold on an on-line subscription form. Documents that score at or above a preset threshold are candidates for delivery through Liberty Mutual's internal E-mail systems or for inclusion in an on-line conference through a groupware facility. E-mail recipients are determined by their personal interest threshold. All documents are maintained in a retrospective database for a limited period of time. IS retains material selected from the news wire for about three weeks and is looking at other mechanisms for storing data for longer periods of time.

As users become more sophisticated, the ability to write content-based queries against the information accumulated from the news wires will become critical. IS expects some subscribers will never need more than basic text retrieval. However, others will want to develop sophisticated queries. IS expects that the system's Macintosh front end will handle the needs of both the casual and "power" user. The long-term goal of IS is to have a range of tools to satisfy the needs of the passive, casual, and power user.

IS carefully selected the first information watch topics to capture the interest of the widest group of personnel in Liberty Mutual. The managed care watch supports the corporation's emphasis on developments in health care cost control. The technology watch meets the needs of Liberty Mutual's widespread information systems community for information on major industry players, start-ups, current products, and future trends.

R&D enlisted IS management to brief senior executives on the potential of the information refinery in terms of their business portfolios and personal information needs. All internal mail subscribers received a brochure that described the information refinery. And, as the system went on-line, R&D's client support services staff members—who are liaisons to the business application development units—hosted presentations with their representative organizations to familiarize them with the system and to develop plans for marketing the information refinery to the end-user community. Almost immediately, R&D received requests for new watches. Within a month, one business unit had taken responsibility for developing and maintaining a life insurance InfoWatch.

The working framework for the information refinery is now in place. However, R&D has several other issues it must address. One is the need to cultivate pockets of query-writing expertise throughout the enterprise so that when the query tool(s) are ready, their expertise can be leveraged. The second is the question of charge-back mechanisms. At this point, R&D has underwritten the information refinery. As new information sources are added, the initiating departments will become responsible for their costs and will, no doubt, want to assess charge-backs to subscribers.

Liberty Mutual found it more challenging than expected to find new information resources for its information refinery. The search for additional news sources found major newspaper publishers lacking in the pricing mechanisms and policies for delivering their publication electronically to an enterprise. Liberty Mutual also found that other interesting sources were only available with a prepackaged text retrieval front end. Because Liberty Mutual is committed to consistent user interfaces, this marketing approach has eliminated several interesting resources for now.

As the information refinery evolved, Liberty Mutual found it needed a broader range of products than a single vendor could deliver. Its ability to implement a solution that could be embraced by the entire organization came in large part from IS's strong commitment to the Macintosh-based client/server architecture, its understanding of the needs and sophistication of the user community, and its ability to work cooperatively with vendors. By following Liberty Mutual's fundamental strategy, R&D could wait for products to evolve that met the organization's long-term plan. This also allowed the R&D group to effectively leverage vendor relationships for the cooperative development of interfaces that made discrete elements of the information refinery work together. Verity, for example, provided tools to build feed handlers for new information sources and Liberty Mutual's IS staff is working with them to provide feedback on the Macintosh-based front end.

In the months following the introduction of the pilot InfoWatches, IS focused its energies on preparing individual business units within Liberty Mutual for full-scale implementation of the information refinery and finalizing contractual details with its information sources for company-wide usage of the information provided. IS reports that virtually every department given an overview of the long-term direction for the information refinery expressed unanimous agreement with the power and simplicity of the approach Liberty Mutual has taken. Individuals have expressed amazement at the simplicity and effectiveness of the single (user interface) window at retrieving a variety of "prescreened" information. Many groups have already initiated lists of sources they would like to see incorporated when the InfoWatch services become generally available.

The information refinery's ability to dynamically create E-mail distribution lists from InfoWatch subscriber lists has proven itself a valuable tool. R&D uses it to see who is using the information refinery and how interested they are in each subject. Other application areas have shown interest in this feature for inclusion in their own systems so R&D has a HyperCard stack which, with minimal modifications, can be used to handle the processing of a subscription form and resulting subscription lists. In time there will be a subscription form of this type for disseminating internal information.

The information refinery is actively pursuing the acquisition of new electronic information from internal documents and major newspapers to research services and regulatory agency filings. These will become more attractive as potential sources develop information feeds. IS is working with Liberty Mutual business units to develop their strategies to exploit the combination of timely information and effective query tools when they become available.

Already Liberty Mutual is targeting the processing of claims, particularly where eyewitness testimony is pivotal. With additional information sources and new interfaces for personalized queries, Liberty Mutual expects the information refinery to become its critical information delivery system for the 1990s and a factor in maintaining its competitive leadership position in the industry.

Text Retrieval Technology

At first glance, the evaluation, selection, and implementation of a text retrieval system seems straightforward. Unlike an imaging system, a text retrieval system is not composed of many separate components. These systems typically are self-contained (i.e., the storage, retrieval, and database functionalities are integrated into a single product). However, after you have made the initiative to further understand the construct of a text retrieval system, the complicated nature of product selection and evaluation becomes apparent. There are several ways that text retrieval can be implemented. The capabilities and characteristics of the system are very much dependent on several basic product features including the system's architecture, the underlying search methodology, the query language and front end, and the query-enhancing toolset. Each of these areas must be evaluated and contrasted to the application's and users' needs and requirements. But in order to do this properly, an appreciation for the role that each component plays and the possible capabilities of each must be appreciated.

Text Retrieval Architecture

Text retrieval solutions are available for the full gambit of architecture from a stand-alone PC to a mainframe host-based system. Since many text retrieval applications use text-only documents, the overall system storage requirements are relatively small. In addition many text retrieval applications support an individual's document collection or one that is easily packaged and distributed to individual users. A prime example is the recent proliferation of CD-ROM document collections, which incorporate text retrieval. On the other hand, the facilitated communications and retrieval provided by text retrieval make it appropriate for multiuser and multidocument collection applications as well.

Stand-Alone. Stand-alone PC-based text retrieval products are numerous and relatively inexpensive. In fact, this is the most common architecture supported by text retrieval products. Although once synonymous with low-level functionality and a lack of sophisticated query-enhancing tools, the single-user stand-alone systems available offer a range of features and capabilities that rival those in the networked and host-based arenas. The most prevalent and only common drawback to these products, aside from the obvious limit to searching in a single document collection, is a low level of customizability. Low-end products are typically integrated with a standard interface and metaphor that cannot be altered to any significant degree. This does not typically present a problem unless the user is accessing several such products for separate document collections. That can present a user with the frustration of having to learn multiple interfaces and searching conventions for each of the products being used.

Host-Based. In a host-based system, multiple users can simultaneously access the document collection and query the system, but all of the documents must reside on a central platform. Furthermore, all processing and index creation is executed on the host machine. The end-user platform functions as a dumb terminal, whether physically one or not.

Historically, text retrieval has been used most in the host-based environment. The reason behind this is as much based in the history of text retrieval as it is in the legacy of most corporate knowledge bases and large document collections. The mainframe was the traditional repository for both of these. It is not surprising that many text retrieval solutions continue to be used on these platforms. At the same time, it is absolutely the case that the host-based market for text retrieval systems has all but dried up. What little activity still exists in this market segment is typically in the host-based mini systems. The mainframe-based systems, though available, are virtually extinct from a new installations reference. What remains is a large installed base of host-based users who are moving slowly to desktop solutions, predominantly in a networked architecture.

The host-based applications that run on mainframes and departmental machines carry a much higher entry-level price in the five to six figure range. That is difficult to justify in all but the largest of text retrieval applications. In contrast desktop solutions can be scaled up as demand and sophistication of the users increase.

Networked. LAN-based text retrieval systems have become the most popular architecture for enterprise users. In fact, although network applications account for only 33 percent of the total text retrieval market, this market segment is the fastest growing, with a 20 percent increase in market size over the last year.

In a network scenario, users will access a single version of the document collection located on one node of the LAN and use the text retrieval engine on that same node to perform the search. Results are downloaded to the local node, where viewing and other operations such as cut and paste may occur. Be aware of text retrieval products that purport to support a client/server architecture but do not provide the ability to handle a distributed document collection or distributed processing. These systems provide customized GUIs on a variety of workstations, but all processing and indexing occurs on the server.

Client/Server. An extension of the LAN-based implementation which accounts for the vast majority of networked installations is that of client/server. In a client/server model the searching and indexing of document collections, as well as the documents themselves, are distributed throughout a network environment. Typically, a significant component of the document collection representing the corporate knowledge base will reside on a server. Individual or departmental document collections reside on remote nodes but may be accessed as part of the overall knowledge base. These schemes can become complex from an administrative standpoint and will require a fair degree of administration to coordinate indices and pointers across even a moderately complex network. Be aware that some text retrieval products do not support distributed indices, although they operate in a client/server environment. In these products, a master index must be created and stored on a central server, or downloaded to each workstation.

The greatest asset of the client/server–based text retrieval model is, however, its scalability. The client/server model provides the ability to tie together individuals, workgroups, and document collections across an enterprise as text retrieval applications grow, while maintaining the integrity of the documents and their access across all users in the enterprise.

On a final note, when designing a distributed text retrieval application keep in mind that requirements for a text retrieval system are not as dramatic as those of an imaging system because text data requires significantly less bandwidth and storage space for transmission and manipulation. As a result basic desktop configurations are adequate for most text retrieval applications. But remember, you may not be using text retrieval exclusively. The text retrieval application might be tied into an imaging system and the user may bring up images as a result of performing a text retrieval query. Keep in mind that we are not dealing with exclusive technologies. Virtually every text retrieval product supports the added capability to attach images to text files via hyperlinks or database indexing. In these cases, the issues regarding image manipulation become relevant to the text retrieval application.

Storage and Indexing

Text retrieval products vary widely in their implementation of an indexing method. At the heart of most text retrieval products, however, is the concept of a database. In some instances, the database is not readily accessible or apparent when you are using the text retrieval system but there is, nevertheless, an underlying database structure.

Text retrieval systems incorporate the full content of documents into the document management system in one of two ways. One approach stores the entire document in the database, a use of BLOB technology. The more common approach, however, stores the name of a document and a pointer to the location of that document on a disk. In this way, the entire document's content is figuratively brought into the record.

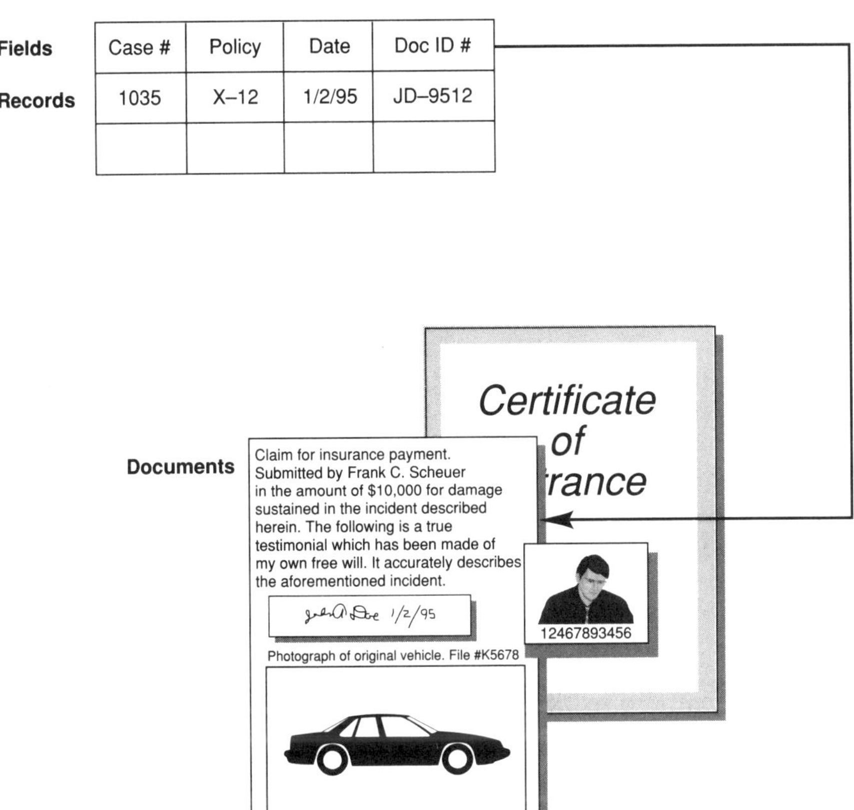

Figure 5-3. Storage. The basic architecture of a text retrieval system is analogous to that of a traditional DBMS. Data is stored or referenced in records which are composed of fields. The major difference is the inclusion of a *text* field. A text field contains the name of a textual document as well as a pointer to the document file. The textual documents can be physically stored in the database or at the operating system level in a native file format. Storing the documents in the database provides greater control and typically a decrease in storage overhead. Storing the documents in their native formats offers greater flexibility for editing and sharing the documents with other applications.

For an example of this approach, refer to Fig. 5-3. In the case illustrated, the data field Doc ID is designated as a text field for unstructured information. What is stored in this field is the name of the document and its respective pointer (the user need only enter the document name, since the pointer is self-creating). In the first record, the Doc ID field contains the document name JD-9512. The text retrieval system automatically builds a pointer to this document and logically makes its contents available to the database. The other fields in this sample database are standard fixed-length fields containing information about the document; in this example, the associated case number, a policy number, and filing date. Combining traditional fixed-length database fields with the document is a feature offered by some products as a way to provide tighter precision over the retrieval process. (For further explanation and discussion of this feature refer to the section on query-enhancing features in this chapter.)

The support of a traditional database record, however, is not universal. In some products, there may be no fields or records to speak of. In these instances what you have is simply an index. You can get to the document through the index but you can't manipulate the index the same way you could a database. Some products support the concept of the database through RDBMS gateways. Under this approach, the text retrieval system does not provide definable database records which support the inclusion of fixed length attributes, but the text "database" can be linked or joined with SQL database records, which can be used to track the documents' attributes.

In integrated text retrieval environments where images and other document types or additional data may be required, you should not underestimate the capabilities of a fully functional database. If you have the ability to work with a fully functional relational database or an hierarchical database, you are probably better off than working with an index structure that you cannot manipulate, that you can't add fields to, or that does not support any sort of structured entry.

Another approach to handling the structure of compound documents and providing increased precision handling through access to document attributes is the use of a tagging language. The most prevalent of these tagging languages, Structured Generalized Markup Language (SGML), is discussed later in this chapter. SGML creates a hybrid database-document environment intended to optimize the benefits of both structured and unstructured document management. The embedded tags permit full-text queries to be limited to specific sections of a document, i.e., footnotes, titles, etc. It is important, however, to realize that not all text retrieval systems will support a tagged document. The support for native SGML files is a design and evaluation consideration that must be confronted. Indeed, the support for any native file type is a feature, not a standard approach, that must be carefully considered.

Some text retrieval systems will convert a document into a proprietary format as part of the loading and indexing of the document. This approach will provide tighter control and security over the documents, but necessitates that the users learn a new editing tool if the documents are to be updated, or the document may be stored twice, in its native format and in the text retrieval proprietary format. The trend in the industry is to support the text files in their native formats, but to provide a metafile view to the text, enabling greater flexibility over the file through the text engine. Text retrieval systems of this type should support a mechanism whereby the tracking of the document is possible so that access without benefit of the text retrieval database is either prohibited or monitored in such a way that the text system is made aware of changes to documents.

The need to control the document and keep abreast of changes to content is crucial to the indexing mechanism of the text retrieval engine. Indeed, the storage issue is very much related to the manner in which the document's contents are indexed by the system for subsequent retrieval. But, before we take an in-depth view of the methods and role that indexing plays in a text retrieval system, it is necessary that you first understand the process of querying the system.

Text Retrieval Query

The query capability of a text retrieval system is a diverse area of concern. There are multiple ways in which a query can be submitted, enhanced, or modified. These various approaches to embellishing a text query (i.e., wildcarding of search terms, heuristic associations, synonym tracking, stemming, etc.) are discussed later in this chapter. At this point, what you need to focus your attention on is the most fundamental capability of text querying, the boolean-based word search.

Once documents are indexed, using whatever indexing approach is available (see the section on indexing methods in this chapter for more detail), content-based searches can be performed. Searches are based on queries submitted. Currently, there are no standards regarding query languages or facilities. (Note that several query language standards, i.e., CCL, Z39.50, Z39.58, SFQL, and CD-RDx, have been proposed and are discussed in the section of this chapter on related standards.) Each text retrieval product has integrated a proprietary query language and/or set of query tools. These languages and tools fall into four basic categories: syntactical boolean-based word query, query by forms, query by example, and natural language query.

Boolean Logic

Although many systems provide front ends and alternatives to boolean word-based query, every text retrieval product supports this basic approach. Boolean-based queries represent the simplest and most basic way to pose a content-based question. In this approach, in its simplest form, queries are based on an equals expression. For example:

FIND RESEARCH.DOC = STRATEGY

This translates to: "Find all the documents within the text field called RESEARCH.DOC that contain the term STRATEGY."

Because these query languages are based on boolean logic, the boolean operators AND, OR, and NOT can be used to link search terms and provide another layer of intelligence in the query. For example:

FIND RESEARCH.DOC = STRATEGY OR DIRECTION AND STRUCTURE

This translates to: "Find all the documents within the text field called RESEARCH.DOC that contain the term STRATEGY or the term DIRECTION and also contain the term STRUCTURE."

Use of the boolean operators is an elementary approach to providing control over precision and recall. The OR operator is a recall-enhancing tool by providing the ability to link several terms together as a single means of identifying relevant documents. The AND operator is a precision-enhancing tool by requiring the presence of more than one word in a document in order for it to be considered relevant.

Recognizing the power of the AND operator and its particular application within a text retrieval environment, the great majority of text retrieval products embellish the AND operator through the support of proximity operators. Proximity is a form of boolean AND linking which allows further control of precision by specifying the proximity of two linked terms. In other words, words must be located within the same paragraph or sentence or within a given number of words to one another. For example:

FIND RESEARCH.DOC = AUTOMOBILE wS ACCIDENT

This translates to: "Find all the documents within the text field called RESEARCH.DOC that contain the term AUTOMOBILE and the word ACCIDENT within the same sentence," and

FIND RESEARCH.DOC = AUTOMOBILE w4 ACCIDENT

This translates to: "Find all the documents within the text field called RESEARCH.DOC that contain the term AUTOMOBILE within four words of the term ACCIDENT."

On this point, perhaps a point of realism is necessary. Although the concept of word-level proximity is very powerful, in practice, most users do not employ this feature. The reason is simple: most of us, end users, researchers, and authors alike, do not think or organize thoughts in terms of word-level proximity. The constructs of paragraphs and sentences, however, are commonly used structures for organizing thoughts and ideas in the written document. Therefore, the paragraph and sentence operators are used almost to the exclusion of word-level proximity. There is one key exception to this fact, which represents a particular usage of word-level proximity, known in many products as *phrase searching*. Phrase searching is a form of word-level proximity in which adjacency (i.e., each word one word away from the next) is implied. Because we do often write in terms of particular phrases, this feature of the text retrieval system is frequently used. The query languages, however, typically do not require that the end user specify the query using multiple "within one word" operators, but recognize the phrase in its "native format." For example:

FIND RESEARCH.DOC = AUTOMOBILE ACCIDENT OCCURRED DURING RUSH HOUR

This translates to: "Find all the documents within the text field called RESEARCH.DOC that contain the word AUTOMOBILE immediately followed by the word ACCIDENT immediately followed by the word OCCURRED immediately followed by the word DURING immediately followed by the word RUSH immediately followed by the word HOUR."

One last point of realism in the discussion of boolean logic: Most text retrieval products support the concept of nested boolean expressions. Many vendors will boast of their product's ability to support up to 16 layers of boolean logic in a single query statement. But, beware—market research has indicated that most users of such systems rarely use this capability. Boolean statements can become unwieldy very quickly. Therefore, most users usually contain themselves to a single query statement consisting of no more than three words and two operators.

Query by Forms

For those that find the construction of a boolean query cumbersome, several vendors have introduced the concept of *query by forms* (QBF). QBF can take one of two forms. One approach is similar to a data entry screen that might be used to populate a database. The names of the fields in the database records are displayed with space for data entry adjacent to each field name, including the name of a document or text field. By providing values for any or all of the fields, the user can construct a boolean-based query.

The other approach is a bit more interactive. When invoked, the QBF facility will prompt the user for a term or phrase. Upon completion of the input, the facility will invoke a pop-up window that prompts for the selection of one of the boolean operators (AND, OR, NOT). The user's selection of an operator will prompt the user for another term or phrase. This interaction will continue until the user enters NO RESPONSE to any prompt. The resulting logic will be constructed into a boolean word-based query and executed. Although this approach can assist the user who is leery of constructing a query from scratch, the burden is still placed on the user to understand and properly utilize the logic of the boolean operators.

Query by Example

Often confused with QBF is a much more powerful feature available in far fewer products, *query by example* (QBE). QBE supports situations in which you are not sure exactly what you are looking for, or how to express your thoughts, but if you should ever encounter the idea, you would recognize it (i.e., "I'll know it when I see it."). An acronym has been coined in this industry to express the "I'll know it when I see it" phenomena: IKIWISI.

QBE is an interactive tool. Typically, the use of QBE is preceded by another form of query, a word-based query, which will identify a subset of candidate documents. QBE removes from the user the burden of having to construct "the best" query. The premise is that the results of a vague query should produce a document that contains text that will more succinctly and powerfully express the concept that the user is trying to hone in on. Thus, if such text is found in any of the documents that are retrieved from an initial query, that text can be highlighted and submitted as a secondary query by the user.

There are many ways that this facility can be implemented from the powerful approach used in parallel processing to the cut and paste approach available with many windows-based text retrieval products. Therefore it is important to investigate the methodology behind the facility, not merely ascertain whether a product offers the facility.

The methods of deployment range from Thinking Machines' massively parallel processor approach for text retrieval to the cut and paste facility of a product like ISYS. In the former approach, the submission of an entire document (no limit to the number of pages) serves as an example by which to determine further relevancy in the document collection. This approach can process in excess of 128 GB of information with a 10,000-word example document in less than 10 seconds. In the latter approach, although there is no physical limit to the amount of text that can be extracted and submitted as a query, there is a practical limit due to the nature of the process and the power behind it. Each word in the highlighted text is linked together by OR to submit a secondary query. Thus if the amount of text submitted is lengthy the user could in effect create a boolean query that is unwieldy and could produce undesired results.

In either case, the underlying benefit to QBE is the elimination of the formulation of many iterative queries. All you have to do is point to a document section that is relevant and submit that as the query. The QBE engine will take that information and use it as the next query.

Natural Language Query

A very small number of products make a natural language query front end available. In this approach, there is no query language per se. Rather, the user queries the system in much the same way that they would ask a question of a research librarian. For example, "I am wondering what effect the collapse of the Soviet Union has had on the world economy, particularly that of eastern bloc countries such as Hungary and Czechoslovakia?" is a valid query in a system that supports natural language processing.

The results of such a query, however, are not necessarily predictable. In other words, as is the case with QBE, the manner in which the query is translated into an executable command will differ from product to product. In some cases the query is stripped of "noise words" (i.e., prepositions and conjunctions) and the remainder are linked together by OR. In other cases the query is analyzed using the same algorithms used to analyze document content. Documents with similar "profiles" as the query are considered relevant (for further discussion of this approach see the discussion on document clustering in this chapter).

It should be reiterated that in the case of QBE and natural language query, the means by which the query is interpreted by the search engine is a function of the indexing scheme. Therefore, a clear understanding of the underlying search methodology utilized by the system is critical to the evaluation of not only the text system itself but also its potential to support QBE and natural language query.

Search Methodologies and Indexing Methods

The underlying search methodology utilized by a product is one of the two fundamental and critical aspects of a product that need to be understood before an educated product evaluation can be conducted. The approach used to index the documents and subsequently retrieve them directly impacts the product's ability to handle dynamic document collections, large document collections, and documents prone to input error, and environments in which the end user is not necessarily proficient in research or specific subject matter. The other critical aspect to be considered when evaluating products is the availability of query-enhancing tools. These tools can augment the functionality of a search methodology, but the basic infrastructure of the system cannot be circumvented, since it is tied to the indexing scheme and search methodology. There are six basic search methodologies:

- The inverted index
- Free-text scanning
- *N*-grams
- Pattern recognition
- Document clustering
- Hypertext

Two of these methodologies, the inverted index and document clustering, can be further broken down into subapproaches. The inverted index can be a positional or nonpositional index. Document clustering can be content-based or statistic-based. These eight search methodologies are reviewed separately. Indeed, a clear understanding of the strengths and weaknesses of each approach is important to appreciating text retrieval systems. But you must keep in mind that these methodologies are not mutually exclusive. In several products, more than one approach is used, each to its own benefit.

The Inverted Index

The inverted index is the original approach to text retrieval, and is still the most popular. One might argue that its age and ease of development make it popular among vendors—the inverted index is used either exclusively or as part of an integrated approach by 88 percent of the products available. Among the end-user community, its continued popularity is undoubtedly due to its simplicity. The mechanics of the inverted index are straightforward and easy to understand. People tend to purchase or put their faith in technologies that they understand. This is the case with the inverted index.

This methodology is based on a self-creating index. The index is created by means of an inversion of all the words in all the documents in the database. The words in the index are arranged in alphabetical order. Thus the index is similar to one you might find in the back of a book. (See Fig. 5-4.) The key difference, however, is that the inverted index contains *every* word in *every* document, with one exception: *stopwords.*

In an effort to minimize the size of the index file, words considered inconsequential to the overall meaning of the document and therefore not valuable as query terms are filtered from the index as it is being built. Those products that support the concept of stopwords will typi-

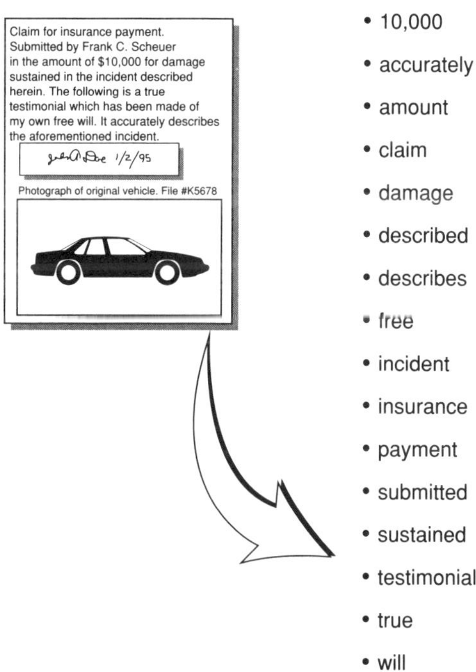

Figure 5-4. Inverted index. The inverted word index is a file containing an alphabetical list of all the words extracted from each document entered into a text retrieval database. These indexes also contain positional information regarding the location of each word in the document collection.

cally ship with a preconfigured stopword file that can be customized (i.e., words can be removed or added to the file). Although there is no standard for stopwords, these files are generally composed of pronouns, conjunctions, prepositions, and various other terms considered to be common in standard written English. For example, a stopword list would contain words such as *a, the, is, so, but, he, she,* and *and.*

In addition to the words themselves, the inverted index tracks the occurrence or location of each word throughout the document collection. The level of granularity to which the locations are tracked will determine whether the system is a positional inverted indexing system or a nonpositional indexing system.

The Positional Inverted Index

This is the more common approach among those products that utilize an inverted index. In a positional index, the exact location of each word is tracked in the index file. Figure 5-5 shows a sample record in a positional inverted index.

In Fig. 5-5, the index record tracks the occurrences of the word *damage.* The record indicates that this term occurs in the database 345 times, and that the first occurrence is located in the document pointed to through the second field (a text field) in the fifth record, the first paragraph within that document, the first sentence within that paragraph, and the sixteenth word within that sentence. Thus, the *exact* location of each word within the database of documents is known within the index file. (If the index term occurs more than once in the database, the multiple occurrences would be tracked using array fields or in a separate related file.)

This level of knowledge tracked within the index leads to the key benefit of this approach, minimal overhead during query processing. Because the exact location of each word is known, there is only one file that has to be opened to satisfy even the most complicated query: the index file. Thus response time is minimally impacted by the size of the document collection and the number of users querying the system. Because all of the information necessary to satisfy any query, even those involving word-level proximity and phrase searching, is all contained in the index, the system need never go back to the documents to satisfy a query.

This index, however, is also the major drawback to this methodology. There is significant storage and processing overhead associated with the positional index itself. Positional indexes range in size from 30 to 150 percent of the size of the document files. The use of stopwords and the application of various sophisticated compression tech-

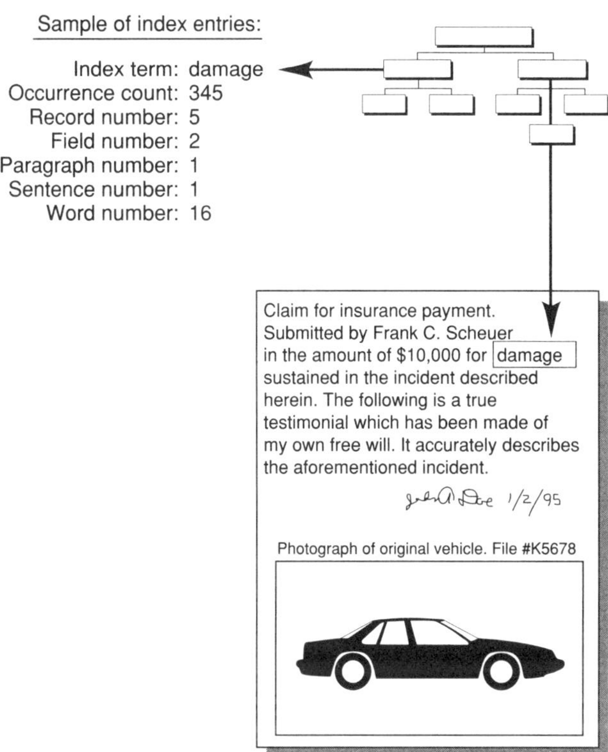

Figure 5-5. Sample record in a positional inverted index.

niques (see the section on associated standards later in this chapter) are responsible for minimizing the size of these files, and the discrepancy in size.

But the index size is only half the story. Because it is the index that is searched and not the documents themselves, a query result is only as up-to-date as the index. Therefore, in most applications, every time the text in the database is modified (i.e., additions, edits, and deletions) the index must be rebuilt to reflect the change in the text. The approach and algorithms used by individual products to maintain the positional index vary. The speed and timeliness of the index updating routine are issues in situations where the text is dynamic and query results must be current. Positional inverted indexes which cannot be practically maintained in real time are best suited for large databases where the text is not subject to frequent modification, or in situations where users do not require up-to-the-second accuracy. In these cases, the index can be rebuilt at given intervals (for example, each day at

12:00 A.M. and 12:00 P.M.). Thus query results will never be more than 12 hours out of sync.

Nonpositional Inverted Index

The nonpositional inverted index methodology is a variation on the traditional positional inverted index approach. What makes a nonpositional index different from a positional index is the level of granularity to which word locations are tracked in the index. Nonpositional indexes maintain term locations only to the document level. The exact location of the index term within the specified document is not tracked. Figure 5-6 shows a sample index record in a nonpositional index. In this example, the index record tracks the occurrences of the word *damage*. The record indicates that this term occurs in the database 345 times, and that the first occurrence is located in the document pointed to through the second field (a text field) in the fifth record. The posting tracks no other information about the term's occurrences.

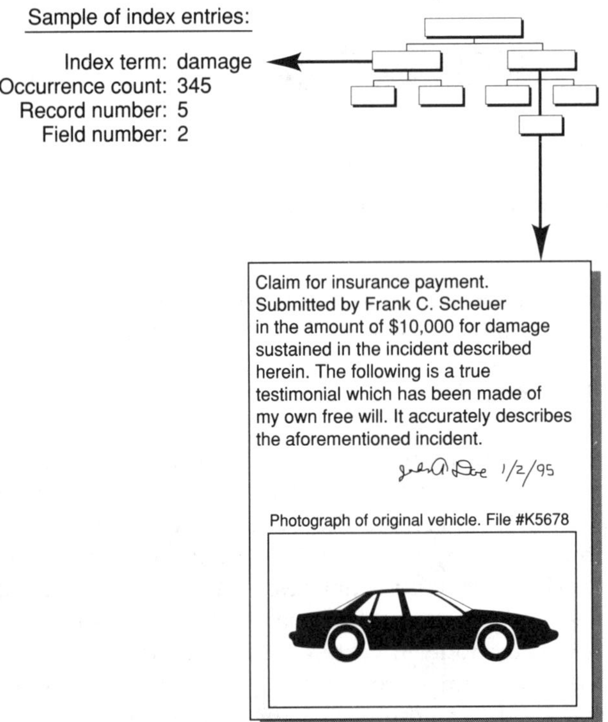

Figure 5-6. Sample record in a nonpositional inverted index.

The advantage of this methodology is the obvious reduction in index overhead. Nonpositional indexes are approximately one-third the size of those in positional indexing systems. The drawback is a slower response time for complex queries (i.e., queries for phrases or those which involve proximity). However, if disk space is a critical issue, nonpositional indexing represents a viable alternative to systems requiring substantial index overhead.

Products which use this methodology typically will support sophisticated queries through the integration of another search methodology. What will differentiate one product from another is how this integration is handled. In each case, however, the secondary search methodology used is free-text scanning.

Free-Text Scan

Free-text scanning is popularly used as a secondary search methodology in products that use nonpositional inverted indices or pattern recognition as primary search methodologies. Free-text scanning is also used as a stand-alone methodology in some products.

The free-text scan approach does not use an index of any kind. In this approach to search and retrieval, each user query causes a sequential read through each document in the database. Products which employ this methodology as the primary or only search methodology integrate a *text array processor* (TAP) to off-load this intensive process from the main CPU. A TAP is a component (typically hardware-based) which accelerates sequential searching of text documents. Thus, the burden of searching is removed from the host CPU. TAPs are configured in a variety of sizes and capabilities.

The benefits to this approach are the absence of the overhead associated with an index, the lack of a need for stopwords (there are none), and the current view to data. Because there is no predetermined knowledge tracked about the documents' contents, each query results in an up-to-date view of the database contents.

The drawback to this approach is the inherent potential for a degradation in response time dependent on the size of the document collection and the number of simultaneous users. Despite the speed of the TAP, because it is a hardware solution there is a practical limit to the processing power. The demands of a large number of simultaneous queries all vying for processing at the same time, and/or a massive text database that requires significant time to be completely processed each time, can slow the search down. It is important that the proper TAP configuration be implemented or response time can be greatly and negatively impacted.

The level of query difficulty will also impact the speed of the system. Generally, the simpler the query logic, the slower the relative response time (compared to an index approach). For example, if a query is searching for one word, the index could locate that word immediately and point to its occurrences. The free-text scan, having no advance knowledge of the word's location would have to search the entire text database. More complex queries, involving rigorous proximity and extensive boolean logic, require significant processing within the index. In these situations the brute force method of a free-text scan can potentially be faster.

When used alone, this methodology is best suited for smaller databases where the text is dynamic and less emphasis is placed on archived information.

N-grams

Products that utilize an *n*-gram approach (sometimes referred to as suffix arrays) construct and use an index as a means to search and retrieval. In an *n*-gram approach, words are not indexed; rather, pieces of words known as *grams* are indexed. In theory, the granularity of the gram is user-controlled (the user can specify that the system build: trigrams, in which case each unique triplet of letters found in the documents is indexed; duograms, in which case the system tracks each unique pair of letters found in the documents; or a unigram, in which case the occurrence of each individual character is tracked in the documents). In practice, however, the vendor has determined the level of granularity of the system, and thus *n*-gram products can be shipped supporting unigrams or trigrams only.

This approach is very similar to the inverted index approach, as an alphabetical listing or index is created and used as the sole means of retrieving documents. The drawbacks to the approach are similar to those of the inverted index—index overhead and maintenance. Although the size of the index of a unigram system is potentially extremely large, vendors of such products utilize state-of-the-art compression algorithms which keep the index size to 50 to 80 percent of the size of the document collection.

It would therefore follow that this approach shares some of the benefits of the inverted index approach, such as minimal impact on retrieval time by number of simultaneous users and size of the database. While this is true, *n*-grams offer other benefits as well. In many cases, the *n*-gram approach will be faster at retrieval. By storing an occurrence table of gram occurrences, the user query terms could be analyzed to determine which of the grams occurs the least within the document collection. The initial lookup would be limited to these less-occurring word fragments and thus have a positive impact on overall processing time.

Further, because character positions rather than full words are tracked, queries can be executed on word fragments. For example, in a medical database, it may be necessary to locate all occurrences of the word root *cardio*. The researcher may not care if the word root is found in terms such as *endocardioplasma, cardioresuscitory,* etc. This functionality is typically thought of in terms of wildcarding the query terms. While other approaches to text retrieval can permit the wildcarding of search terms, they require added functionality and can potentially result in slower response times. When using *n*-grams, wildcarding is an innate feature with no degradation in response times.

The tracking of grams also makes it possible to perform lexical analysis on the database, rather than word-based retrieval. This functionality falls into the serendipity category of research. For example, queries such as *What is the most frequently occurring phrase that begins with "Read my lips" in this database?* are possible. Similarly, the query *What is the longest phrase that occurs in more than one document?* is also possible. If this query results in the retrieval of several documents with a lengthy common phrase, written by different authors, without cross-reference, then plagiarized text was uncovered, using virtually no effort. The nature and value of this type of lexical analysis is subject to user and application need and imagination.

Pattern Recognition

Pattern recognition is the newest approach to text retrieval. It is similar to the inverted index in that an index is created and used as a means of retrieval, but in this methodology words are not indexed. The words in the file are reduced to their binary representation. A proprietary mathematical algorithm is then applied which calculates the "patterns" that arise from a body of text. These resulting patterns or bit vectors are tracked in the index file along with pointers as to where the patterns can be found in the original documents.

Using this approach, the text queries can be entered in natural language. This is not an add-on feature to the products that support this search methodology, but an innate feature of the methodology itself. (If the user desires, a boolean-based query can be used, but this is not necessary.) The user query is processed using the same algorithms used to process the document text. The patterns that result from the query are compared with those in the index. Those indexed patterns that appear to be similar to the query patterns are used to point to relevant documents.

There are several benefits to this approach. First, because the system reduces the words to bit vectors, processing is usually faster since computers can crunch numbers faster than they can interpret letters. This methodology typically runs under a neural network which further increases the pattern-matching speed. The usage of bit vectors also accounts for relatively small indices. Products that use pattern recognition create indices that range from 7 to 30 percent of the size of the text documents.

Furthermore, because the methodology is not based on exact word matches, it is not handicapped by spelling errors in either the body of the text or the queries. Likewise, the methodology can innately handle subtle differences in word spellings such as in plural and past tense constructions.

While functionality such as this is possible to one degree or another using other search methodologies, in these cases it would require additional query-enhancing tools (discussed later in this chapter). Pattern recognition offers this functionality innately. Pattern recognition is not an exact science. In other words, the retrieval engine is looking for similar patterns—not exact hits. Words that are similar in construct can result in the same pattern. Thus, pattern recognition is tolerant of spelling errors. If a user is not sure how to spell the query term *chrysanthemum,* this is not a problem. A best guess should find documents that contain the word. This is sometimes referred to as fuzzy searching.

This benefit works in reverse as well. If the documents contain misspelled words, a correctly spelled query term should still uncover the document. This can become important when OCR is a major source of input into the system (OCR input is prone to error). Where approaches such as the inverted index and free-text scan search for exact matches between query terms and indexed terms, pattern matching requires only a similarity between them. This benefit, however, can also be considered a drawback, depending on the orientation of the application and the users.

The ability of the pattern recognition engine to perform fuzzy searches can have a negative impact on the system's precision. Because different words can have similar patterns it is likely that a search for one concept can result in a fair number of documents that have nothing to do with the concept, but demonstrate similar patterns. For example, if conducting research on telephones, it is likely that the system will return documents discussing elephants. Conceptually these two subjects have virtually nothing to do with one another, but telephone and elephant result in patterns that are similar and therefore the search engine sees them as related.

Pattern recognition products can support exact word matching, but in these cases a free-text scan is being used as a secondary search methodology to ensure that exact, and not fuzzy, matches have been found. The pattern recognition methodology is initially used for its speed and efficiency to locate candidate documents. Candidate documents are documents that exhibit a high probability of containing the search terms. Using the alternative methodology, these documents are then searched for the exact word matches.

User-selectable parameters control the balancing of retrieval speed and the degree to which pattern likeness is required. These parameters permit the user to establish evaluation guidelines such as the minimum number of hits per document necessary to qualify as a candidate and misspelling and word-order tolerance levels. But it must be understood that the higher the precision demands, the slower the response time. This trade-off of response time for better precision, and the general sacrifice of precision for better recall, is a basic design requirement that users have to contend with.

On a final note, perhaps the most fascinating feature of this methodology is its potential as a content-based retrieval tool for data types other than text. Because retrieval is not based on words, but bit vectors, it is logical to conceive that any data type (images, voice, video) can be reduced to its binary representation, processed with the proper algorithms to produce the patterns that represent the data. Thus, queries comprising a video clip, a sound bite, or an image can be reduced to patterns and matched with similar patterns of respective-type stored files.

Document Clustering

Document clustering text retrieval permits more sophisticated queries by incorporating intelligence into the search and retrieval process. Using this methodology, retrieval can be based on ideas and concepts, not simply words and phrases. While vague, conceptual queries are supported in products that utilize other search methodologies, this is only through the integration of query-enhancing tools (see the section on precision and recall enhancement in this chapter for further discussion). Conceptual querying is an innate feature of the document clustering methodology.

Key to this approach is the means by which the intelligence is built into the search engine. There are two approaches: concept-based clustering and statistic-based clustering. Before the differences between the two approaches are discussed, however, the premise on which both approaches are based can be examined. In clustering, an index is built. This index is built at the document level, as opposed to the word level, pattern level, or gram level. The overall contents of a document are examined, using the intelligence of the system, and a resulting index value for the document is created. In this way a single word or phrase does not denote the presence of a topic in a document, but rather the cooccurrence of several words gives way to the overall document meaning. For example, taken individually the terms *Dallas, limousine, 1964,* and *motorcade* can lead to any number of ideas. But when viewed collectively, it is reasonable for a reader to conclude that the document is addressing the Kennedy assassination. Thus, if a query is submitted for information on the Kennedy assassination, despite the fact that a document does not contain the word *assassination* or the word *Kennedy,* the document should be retrieved if the four terms listed above are used collectively throughout a document. Inference reasoning is used.

Using the intelligence of the system as the basis for inference, each document is indexed against any number of subject matters. Any number of subjects can be tracked. The determination of the subjects begins a process by which a virtual storage area is constructed (see Fig. 5-7 for further clarification). The virtual storage area is an *n*-dimensional space, where the number of dimensions is directly related to the number of subjects identified, one dimension for each subject.

The documents in the database are parsed, extracting words and phrases. Using proprietary mathematical analysis (closely linked with the intelligence imposed on the system), which uses the documents' vocabulary and word positioning, the subject contents of the documents are determined. The documents are placed into the multidimensional virtual storage space based on their relationship to the different subject matters. Because a document can pertain to several subjects, it is possible that a document would be placed in several positions within the virtual space.

Once the documents are placed in the storage area, users can query the database via a natural, conversational language. Although natural language query can be integrated into other search methodologies, it is an innate feature in this approach. Thus, questions such as: *What is available concerning illegal activities on Wall Street?* and *Hey, I am wondering what effect the collapse of the Soviet Union is having on the economy of the United States?* are valid queries. The result would not be a succinct directed response, but a listing of documents that appear to contain the information needed to answer the question. Furthermore, the

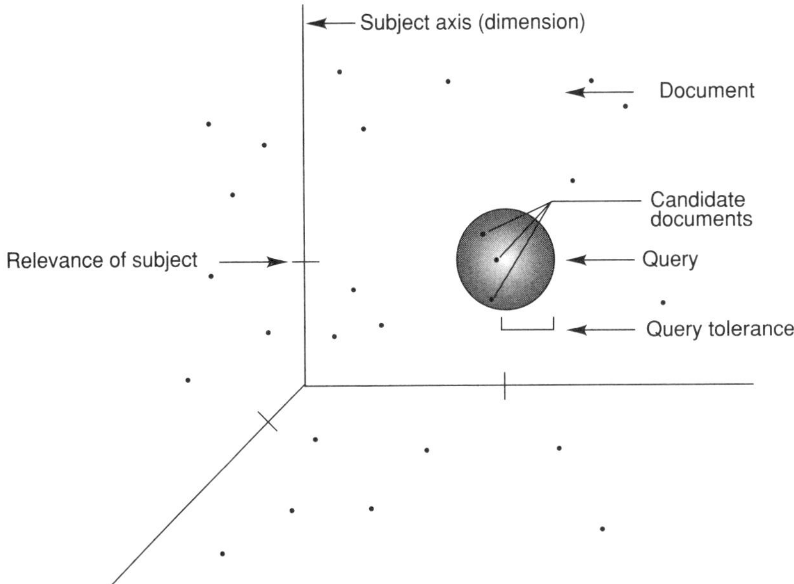

Figure 5-7. Document clustering. In document clustering, a vector, or subject profile, is created for each document. These vectors create a virtual storage area of *n* dimensions. Because the storage area is virtual, based in mathematics, the dimensions of the area are infinite. The virtual dimensions are determined by the number of subject matters that are present. Each subject or interest category creates a vector or dimension within this storage area. The documents' subject profiles are used to position the documents within the storage area. In a similar fashion, user queries are also translated into a vector which positions the query in the storage area. Retrieval is based on the proximity of a document's location to the user query.

retrieved documents would be listed in the order of relevancy, so that the first document listed appears to address the question most aggressively, the second document is the second best resource for this subject, etc. Note that the ability to rank retrieved documents by relevancy is a feature that can be integrated into other search methodologies, but is an innate feature in document clustering approaches. (For further discussion of relevancy ranking and its availability in other approaches, refer to the section on precision and recall enhancement in this chapter.)

The associating of documents to a query is accomplished by processing the query using the same logic and methodologies that were applied to the documents. Thus, a query is given an index value which positions it in the virtual document storage area that was created. The point at which the query is situated is considered to represent the essence of the subject matter of interest. It is known as the query centroid. Because the figurative positioning of documents within this area is based on subject content, those documents that occupy the same general area as the query centroid are assumed to pertain to the same subjects as the query. Therefore, a virtual sphere is formed around the query centroid. All documents whose index values fall within the sphere are returned as relevant. Thus a query would retrieve all relevant documents, not just those containing the query terms. By controlling the radius of the sphere formed, users can control the degree of relevancy required in order for a document to be returned.

The benefits to document clustering are fairly obvious. Retrieval is based on concepts, not words. Therefore users are not required to succinctly and precisely describe the subject of interest. Intelligent responses can result from naive queries. The query process is facilitated through the innate support of natural language query. This approach however, should be used only in large (multigigabyte) collections. The ability to distinguish subject matters between documents and similarities among documents is very much related to the amount of documents that can be compared. Response time is not affected by size of the database, and response quality is actually enhanced as the database grows. Thus if your application involves a large collection of documents, this feature of clustering would also be considered a benefit. These benefits are true for both approaches to document clustering. The drawbacks, however, are related to the specific approach taken.

Concept-Based Clustering

The concept-based clustering approach requires that a great investment in time and personnel resources be expended up front. This approach to text retrieval is literally useless until a staff of highly knowledgeable individuals take the initiative to instill knowledge into the search-retrieval clustering engine. Under this approach, installation includes a process by which the system is "taught" how words relate to subjects. By using a numeric rating scale (0 = not relevant, 9 = extremely relevant), subject matter experts rate the words and phrases that are likely to appear in the text database against the subject matters of interest in the database. For example, the phrase *insider trading* would be closely related to the subject *stock market* but not relevant to *entertainment.*

Because these systems work on large document collections the number of intersections of words and phrases to subjects can get quite large. It is likely that each expert would expend from two to three days going through this process. (It is advisable to have more than one person go through this exercise and to merge the results. If only one person's input is used to build the intelligence of the system, it is likely that the system will reflect personal prejudices and perspectives. These can be eliminated or tempered through multiperson input.) After this exercise is complete, the system must interpret the input to build the virtual storage area. This process can take another 24 hours. Once completed, the documents must be processed and indexed, utilizing the acquired knowledge. Depending on the size of the data collection this can take another two to three days.

The processing time associated with building this clustering environment makes it impractical to update the logic or knowledge of the system on a regular basis. Every change in the system's knowledge will require a rebuilding of the storage area and a reindexing of the document collection. Therefore, although this approach can handle a document collection that covers a broad range of topics and issues, the overall perspective on these issues and the way language relates to these issues should be fairly static, or this is not a valid approach to text retrieval.

Statistic-Based Clustering

Statistic-based clustering also requires a significant investment in time as part of the installation process, but does not require personnel investment. The knowledge used by the system to build the document clusters is gleaned from the document collection itself. Utilizing an inverse-frequency-weighting algorithm and statistical analysis of the word occurrences and cooccurrences, the relationships of words, phrases, and concepts are determined. These relationships, based on statistics, embody the knowledge base used as the foundation of the resulting database. The biggest drawback specifically associated with this approach to clustering is the reliance on statistics. Certain individuals are not comfortable surrendering the definition of their document database's knowledge base to a process that is as arbitrary as statistical analysis.

Hypertext

The final approach to text retrieval, hypertext, is radically different from the other methodologies discussed. Therefore, in order to fully appreciate this technology some background discussion is necessary.

The other approaches to text retrieval are modeled after the queued memory process. Queued memory processing requires prompting from an outside source to stimulate the retrieval of information. In the case of a text retrieval system, this prompting takes the form of a query. Hypertext, on the other hand, is based on the associative memory process. Associative memory is triggered by an internal thought or idea. The instance of one thought triggers another thought or idea. There are many examples of manual hypertext systems such as an encyclopedia: the information on one subject will undoubtedly contain references to other related topics, indicating where to get more information on these particular topics. Almost instinctively, humans will mentally build associations between separate pieces of information to embellish our understanding of either or both information sources, or as a means to help prompt memory processing.

Hypertext is a software-based approach to capture these associations and make them available as part of the research process. Hypertext is as much concerned with the interrelationships between documents as it is with the subject matter of each document. These systems allow for the departure from linear organization of flat text files. Text can be stored and searched in a hierarchical or network fashion. Hypertext offers users of electronic document management systems functionality unique to its architecture and mechanics. Hypertext permits browsing of a document collection. This interactive nature places hypertext on a plane similar to interactive human communication. Its revolutionary approach to information organization, access, and authoring has stimulated much new interest in the role computers can play in information management and instruction. However, hypertext is often misunderstood or mistaken for an expert system or full-text retrieval system. While there are similarities, and the integration of the technologies is possible, hypertext is neither of these systems.

Hypertext is not a very useful tool for the initial research endeavor. It offers little assistance in locating a logical starting point within a document collection. It is, however, an excellent navigational tool that provides guidance and insight as to where else in a document collection a user may find relevant or collaborative information, once positioned within the document collection. For that reason, hypertext very often is used in conjunction with one of the other search methodologies. The "traditional' approach is used as a primary research tool to establish a foundation of documents. Hypertext is subsequently used to further research without requiring subsequent queries. (In practice hypertext functionality is used for a variety of other purposes such as context-sensitive help, and a facilitated means for supporting compound documents. These facets of the technology will be discussed later in this section.)

Surprisingly enough, hypertext as a concept is over half a century old. The idea for a desk-based device that would store the information of an entire library and allow for quick intuitive retrieval is credited to Vannevar Bush in 1938. In 1945, the former science advisor to President Roosevelt presented his idea to the public. In an article in the *Atlantic Monthly,* Bush christened his idea *Memex* (MEMory EXtender system).[4]

The Memex was described as a private filing system, emphasizing a personal nature and scope. Storage of all types of information was to be accomplished through microphotography. A mechanism of some type would allow the linkage of any two pieces of information. Bush called this concept an information trail. The Memex never reached its full realization. Technology at the time could not support Bush's design, but there is no doubt among today's hypertext proponents that it was Bush's vision that gave birth to hypertext.

It was not until 1963 that further work was performed in this area of information management. Influenced by Bush's work, Doug Englebart (inventor of the mouse) designed a computer system called NLS (*oN Line System*), which embodied many hypertext concepts. This system allowed scientists at Stanford Research Institute to link documents on a mainframe. The system evolved over time and is now used and marketed by McDonnell Douglas as *Augment.* At about the same time, Ted Nelson and Andy Van Dam began work on a system named *Xanadu.* While Bush is credited with the idea for hypertext, it is Nelson who is credited with coining the term. Speaking at an annual ASIS meeting, Nelson spoke of using Bush's concept in Xanadu (then a term project for a graduate course at Harvard University), and referred to the technology as *hypertext.*

Though additional work with hypertext continued into the 1970s, it was not until the late 1980s that the technology began to receive widespread recognition and interest. The advent of related technology, such as powerful workstations, high-resolution screens with windowing capabilities, and storage media capable of holding vast amounts of text and images in little space (CD-ROM, WORMs), gave the technology the platform it needed. Furthermore, the expansion of hypertext into the realm of hypermedia has also spurred its acceptance.

Hypermedia systems provide hypertext-like linking macros that will link any digital data file to any other digital datafile. Thus, a block of text could link to an on-line image, which could link to a voice file, etc. Hypermedia links, for example, are a popular approach to supporting document images from within a text retrieval system. Although the term *hypertext* specifically refers to the linking of a text block to another text block, the term is more widely used today and

[4]"As We Think," *Atlantic Monthly* (1945).

has come to embody the concept of linking of any data types in this fashion. The term hypertext will be used in this book, but it will encompass the definition of the hypermedia system as well.

Components of Hypertext

The Semantic Network

At the core of a hypertext system is the semantic network, or mechanism by which the elements in the system are integrated. These semantic networks can either be independent of the system, or embedded within the system. In independent networks, the units of information in the network are tagged with concepts represented by terms. Ties can be built between sections of one document to sections of another document. In the embedded network, the ties are from one document to another. It may be necessary for a user to read through several blocks of text before discovering why one document was linked to the other.

Nodes

Each piece of information, whether textual, graphical, or audible, is stored in a *node* (also known as a clump, chunk, card, or pad). A node is the basic unit of storage in the system, analogous to a field in a traditional database. The node is usually displayed in a window with its identifying title. A node can hold any amount of text, images, computer-generated graphics, on-line video, and sound.

Links

Nodes are interconnected via *links.* Links are the pathways from one node to another. Links are typically activated by the click of a mouse or a few simple keystrokes. The process of enacting a link to move from one source of information to another is called *traversing.* Traversing is accomplished through macros. These macros are generated automatically by the hypertext system, based on the author's input. Links can be used to connect two separate nodes (a similarity link) or two locations within the same node (a contiguity link). Links can also be used to connect annotations to a document, similar to an electronic Post-it, and to provide organizational information, such as where the node fits within a large document.

Links usually originate from a single point, as from a word or sentence. This point is called a *link reference.* The other end of the link, the destination, is known as a *link referent.* Link referents can be nodes, a region of text within the node, or a portion of another node.

Buttons

Links are represented by *buttons.* A button is a block of highlighted text or, more commonly, an icon that usually identifies whether it is a source or destination button. Optionally, it can contain the title of the destination and the link type it represents. By clicking on a button a user traverses a link.

The Browser

The *browser* is a useful tool, but is not a standard feature in hypertext systems. The browser aids with navigation. In some systems the browser is a node. This node will contain a graphical depiction of the network. The depicted nodes in the browser act as buttons, traversing the user to that node. The browser can be called from a node, displaying just that part of the network centered around the current node. Some browsers will display the current path of links and nodes used to arrive at the current location in the network.

Paths and Filters

Two other tools used to assist in the navigation of the network are *paths* and *filters* (also known as stacks or pads). Paths are hardcoded default links through the database. The buttons here are automatic, and at times invisible. When a path is followed, the user surrenders navigational control and the ability to control the research flow to the original author of the system. Filters suppress detail from a node's contents. They allow the user to quickly scan the nodes in a network for possible relevancy.

Advanced Features

Other features which have yet to become established as standards in hypertext products are sensitive paths, automatic button and link creation, and live links.

Sensitive Paths

Sensitive paths are semiautomatic paths that are linked to a user's login, or to user responses. These paths will automatically direct the user's session in certain situations. For example, in an ecological database, depending on the user's personal bioregion, different examples will be used. A single medical reference book could be created that will serve the needs of doctors, administrators, nurses, and patients alike. Based on responses to key questions the system can dynamically determine a user's level of expertise and create an appropriate version of any manual or text. Certain information is shared, but each user level is automatically linked to the appropriate technical detail and examples. In this way virtual chapters are created by organizing a common set of nodes into a variety of paths, each creating a unique view into the database, tailored to a user's needs.

Automatic Button and Link Creation

A number of products have begun implementing the ability to accept formatted text (text produced in a word processor) and output a hypertext version of the information. Each of these products implements this functionality somewhat differently, but the availability of such functionality has been a large step forward for the practicality of hypertext. Each of these systems provides the ability for the user to modify the system-generated links and buttons. Typically, the more information required from the user prior to processing the data, the less user modification is necessary.

Live Links

Live links support the integration of nodes that contain dynamic file types (i.e., an active spreadsheet). Each time a link to the file is traversed, the file will automatically update itself with the latest revision.

Hypertext Architectures

Hypertext can be used to create one of three types of document architecture or a hybrid of all three. These architectures are *hierarchical intradocument, networked intradocument,* and *networked interdocument.*

Hierarchical Intradocument

This is conceptually the simplest hypertext architecture. In this model, a document is decomposed into layers of detail. The reader of the document can determine the level of detail that they wish to read on any given subject within the document. (See Fig. 5-8.) This is a popular approach for commercial publishers of catalogs and other such publications.

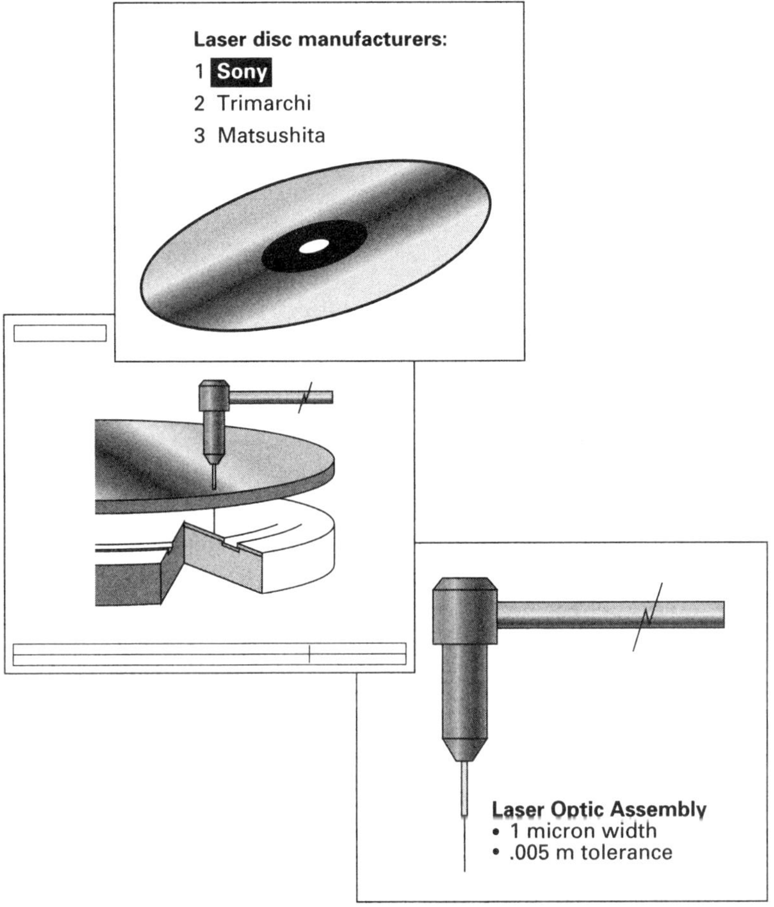

Figure 5-8. Text retrieval. Using a hypertext system, levels of detail of information on a subject can be linked in a hierarchy. In this example the user begins with a listing of laser disc manufacturers. By selecting Sony, the user is provided a patent drawing of the manufacturer's WORM system. A click on the read-write mechanism invokes a detailed drawing of this component.

Networked Intradocument

Intradocument linking can also be used to create a network of cross-references, rather than levels of detail. This approach is very popular in the electronic publishing arena at large. It is especially useful in publications that require several sections to be processed by a user in order for a complete understanding of a subject to be attained. Electronic encyclopedias and "how to" manuals are good examples of this approach to hypertext. (See Fig. 5-9 for an example of this approach.)

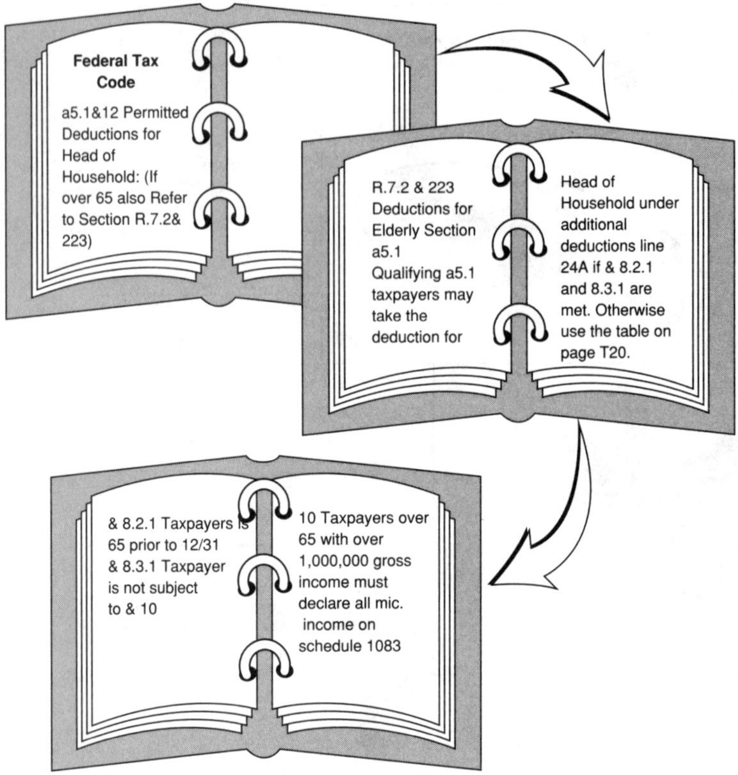

Figure 5-9. Hypertext. Intradocument networked links can be used to present information typically viewed as a single book, in a manner most conducive to the reader. By utilizing filters and paths, the concept of customized virtual publications that are specifically designed to support each reader type can be created without the effort of physically creating the separate versions. In this example, the user of an on-line *Federal Tax Code* is dynamically linked to three separate sections of the code. By utilizing self-executing links and sensitive paths, the author can create multiple versions of a document.

Networked Interdocument

This is the original architecture under which hypertext was envisioned. In this approach, the links are used to create a network similar to that described in intradocument linking, but the links connect to data sources outside the originating document. The relationship between documents in an entire library can be tracked and made available as part of the retrieval process. The level of granularity supported by the links is important in these situations. The author may want to link a particular paragraph of one document to another paragraph in another document. If the links are supported at the document level, the link would have to be from document to document. The logic of the link (the text in a particular paragraph) may elude the reader who is not apt to read through paragraphs in an attempt to determine why they were linked to this document. (See Fig. 5-10 for an example of this approach.)

Problems

There is a significant level of awareness and acceptance of hypertext. Various types of authors are realizing its benefits, from providing context-sensitive help to producing effective electronic CD-ROM publications. Despite this, however, evaluators and authors of hypertext systems must remember that lack of careful attention to the design and creation of the network can create an almost insurmountable problem in a hypertext system—a meaningless web of obscure connections,

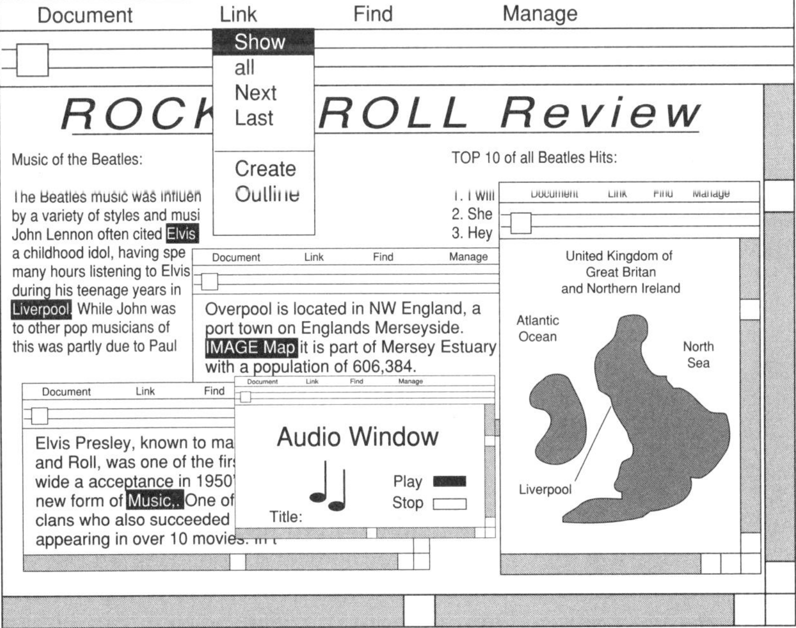

Figure 5-10. Example of a hypertext system using windows representing multiple nodes. In this example, the reader of the Rock and Roll Review article has traversed two buttons, leading to hypermedia information on the city of Liverpool and the life and music of Elvis Presley. Typically, the user would have closed the windows associated with one path—starting another path.

labyrinthine in complexity. Unfortunately, hypertext's greatest feature, ease of use, is also its greatest danger. In most systems, creating links is not difficult. A novice hypertext author may be tempted to zealously create many different nodes of many different types and connect them with a variety of links, only to find that navigation through the system is almost impossible.

If the author-designer will not be the only user of the system, care must also be taken to ensure that the end result is an objective collection of related information, useful and meaningful to just about anyone. Authors must be reeducated on writing skills and approaches. Thinking in terms of sequential processing, or designing documents that are meant to be read from page one to page "end," must be challenged. The author-designer must be able to think in terms of information "chunks" and be able to objectively determine the various ways these chunks should be linked together.

Readers or users of these systems are offered little support as well. There is as yet no standard, proposed or otherwise, regarding the browsing facility. Many browsers can be as confusing as the network itself. Many products have no browser. In these cases it is likely that a user can get "lost in hyperspace." As comical as this may sound, becoming lost in a web of cross-references is not a funny or pleasant experience. I can recall one evening doing research on the history of the presidency of the United States, using a hypertext library. After only one-half hour of investigation, I found myself reading a dissertation on Kentucky blue grass farming. What does this have to do with the presidency? Indeed, what did it? I could not figure it out, and in the absence of a browser of any sort was compelled to log off the system and start the research process over. A waste of time—yes. But it also caused a high level of anxiety and a sense of disorientation that did not easily go away.

In cases where a browser is available, there is no guarantee that it will be helpful in all cases. Some browsers, for example, cannot communicate closure to the user. Closure is knowing when you are near the end of a path or stream of thought. With a book, closure is simple to perceive, but in a hypertext system, without the appropriate set of tools, the user may not be aware of nearing the end of a thought stream. Often, knowledge of the path end can make more sense of the path and provide a sense of security to the user. Depending on the system, this problem can be compounded by the fact that some systems require that each node contain no more information than can be displayed on a single screen. Other products allow more than a screen's worth of text to be stored in a single node, and provide scrolling for viewing the text.

There is no standard for beginning a session in hypertext. Some systems rely strictly on a browser, while other systems have incorporated key word searching or full-text retrieval. The ability to begin a session with a full-text or key word search is useful when users begin with only a vague idea of what they are looking for and do not know where to find it.

Regardless of the type of system and the tools available, the author of a hypertext system must realize that he or she is in control and must drive the system. Hypertext systems aid in information organization, adding clarity and speed to retrieval. They do not determine the content of the system, the link types to be used, or the organization of the network. If the author's associative talents are not strong and obscure or meaningless links are created, the end result could be a useless system. For the user of a hypertext system, similar talents are needed. Deciding where to begin the research and which path to follow is up to the user. If poor decisions are made and you lead yourself down obscure paths, it may become difficult to find any valuable information at all.

In systems that do not support rich text (i.e., different fonts, indenting, underlining, etc.), the conversion of text into hypertext can result in a loss of structure. Loss of structure has an adverse effect on readability. In some systems, windowing is not available. This too has an adverse effect on reading comprehension. Though Vannevar Bush himself spoke of the importance of simultaneously viewing several linked nodes at the time of creation and retrieval, there is no universal agreement that windowing capability should be required in hypertext, though most commercial implementations support this model.

Conceived initially as a single-user system, hypertext has been easily converted to a multireader environment, but has had difficulty supporting the multiauthoring environment. Not every product supports this concept. Cases in which multiple authoring is allowed require a higher level of control and design criteria. Lack of standards, design rules, and expertise render the creation of a large comprehensive multiuser hypertext system a time-consuming process requiring careful planning. Do not be taken in by the ease and speed by which a small demo system of 6 to 12 nodes and 4 to 5 links can be accomplished. At the same time you should keep in mind that not every hypertext system needs to be all things to all users. For example, the porting of a single manual to hypertext format is a much more practical undertaking than converting an entire corporation's library. In considering hypertext applications, time and effort involved should be weighed against the perceived benefits. Hypertext's current problems are not insurmountable. Many applications and products based on this technology do exist and are having a positive effect on the way we use and communicate information.

Precision and Recall Enhancement Tools (Query Enhancing)

Comprehensive and effectual evaluation of text retrieval systems is as much dependent on an appreciation for the query functionality and features of the system as it is on understanding the underlying search methodologies. While the search methodology provides the infrastructure for the retrieval process, functionality provided through technology and tools as simple as term wildcarding, and as sophisticated as conceptual text analysis, provides the flexibility and power necessary to create text retrieval solutions that fit the user environment. It is through the addition of this functionality that text retrieval goes beyond being a fast but ignorant retrieval engine, and becomes an intelligent gatherer and assimilator of information.

Therefore, the availability of add-on tools that work in conjunction with the underlying search methodologies is the second critical area of text retrieval product evaluation. These tools, which if available are typically integrated into the text retrieval product, enhance the ability of the system to either directly or through user interfaces offer better control over precision and recall. This type of control is a critical issue since effectiveness in a text database is not as straightforward as it is in the traditional structured databases. Text retrieval databases inherit the ambiguity, richness, and subtleties of the data type that they track, written free-form language. Thus, the author of a document may describe an event one way, but the researcher executing the query refers to it another way. If a text retrieval system is effective, it will allow the user to locate information relevant to the query, even if the query does not state the search term in exactly the same way as it is contained in the text.

Effective design of a text retrieval system demands that the designer recognize the tools and techniques necessary to deliver a system that addresses all user levels of expertise, while maintaining precision and recall, and conforming to the work environment. In some cases the tools are ubiquitous; in others they are interactive, requiring user decisions and input. The need and application for these tools are dependent on the level of user expertise on the subject matters being queried, the volume of information, and user familiarity with the information.

These tools are discussed singularly, but as is the case with search methodologies, these query-enhancing tools are often deployed in tandem. The tools are discussed in order of complexity of use and level of control provided.

Use of a Traditional Database

As discussed in the storage and indexing section of this chapter, some text retrieval systems provide an integrated traditional database structure. Others provide database functionality through gateways or preestablished links to third-party RDBMS. In either case, the availability of a database provides precision-enhancing control. The premise behind this is that large document collections contain too much information and potential relevancy to adequately differentiate documents. The adjunct of a database enables users to rely upon a degree of structure with which to navigate through a document collection. We are not distilling the document; instead we are providing one more means by which to identify relevancy and conduct investigative research without the burden of excessive filtering by the user.

Examples may be using structured items such as *DATE, AUTHOR, TITLE, SUBJECT,* or *PUBLICATION* for this purpose. Thus, the user can decide on a case-by-case basis whether to use these values to narrow down the candidate collection of documents. As a result, precision is increased. For example, if you were interested in gathering information regarding the effects of radiation on eyesight, your query would be based on words such as RADIATION and EYESIGHT. This type of query could lead to a large collection of documents. But what if you were only interested in information published after 1990, and in a medical journal? If publishing date and document source were tracked in a database, your query could include these qualifications and thus narrow the scope of retrieval, or increase precision. You would be given a more focused view of the document collection, based on your specific criteria. If publishing date and source were not tracked in a database, these discriminating values could not be included in the query. This would more than likely result in a larger document collection that you would have to peruse searching for clues within the body of the text regarding source and date, and thus perform a second-level elimination manually.

Many evaluators of text retrieval often overlook the need for a database. Their attitude is to throw everything into a big bucket (the text database), and then use the text retrieval system to infer relevancy from vast amounts of information. It doesn't work that way. The larger the bucket gets, the more information you put into it, and the harder it becomes for the user to quickly identify relevancy. Structure for navigation purposes is very important and very useful. This is one of the most valuable roles the database can fill.

Numeric and Date Search Capabilities

Although all systems will handle documents that contain numeric and date information, not all will process the information as a number or a date. Some systems will simply skip or ignore numeric characters. Others will treat the numeric characters simply as characters. Systems with support for numbers and dates will not only process the characters but are also intelligent enough to understand that they represent a particular value. Therefore, queries can contain numeric and date ranges; for example, searching for all documents that contain a numeric value between 100 and 130, or referencing a date between January 1, 1992, and February 12, 1993. Thus, the ability to search on numbers and dates is a precision-enhancing tool.

Wildcarding

Wildcarding provides a simple, yet powerful and efficient mechanism for simulating multiple boolean ORs and thus exercises greater control of recall. Wildcarding manifests as a query language feature. The wild card is a reserved character that can be used in place of a single character or any number of characters in a query term. For example, the sign for *at* (@) can be used to replace any single character, while the asterisk (*) is used to replace any number of contiguous characters. Wild cards can be used to eliminate the need to know the exact spelling of a word or as a shorthand means of searching for words with a common root. For example:

```
FIND RESEARCH.DOC = SM@TH
```

The translation is: "Find all the documents within the text field called RESEARCH.DOC that contain any terms beginning with SM and have exactly one character following, followed by TH." Thus the query would result in retrieval of the name SMITH as well as SMYTH. If asterisks were used (following the wild card rules outlined above), the query would have searched for any words that begin with SM and end with TH and have any number of characters in between. Thus in addition to SMITH and SMYTH, words such as SMOOTH would also have been searched for. When embedded in the middle of a text string, the wild card is referred to as a root word wild card.

By placing the wild card at the end of a search term, suffix searching can be accomplished. For example:

FIND RESEARCH.DOC = DISCRIM*

The translation is: "Find all the documents within the text field called RESEARCH.DOC that contain any terms beginning with the prefix DISCRIM." (This would include discrimination, discriminatory, discriminating, etc.)

Prefix wildcarding is accomplished through the placement of the wild card character at the beginning of a query term. For example:

FIND RESEARCH.DOC = *ING

The translation is: "Find all the documents within the text field called RESEARCH.DOC that contain any terms that end with the suffix ING." This query would search and retrieve on words such as running, seeing, believing, doing, etc.

When evaluating wild card capability you should keep two things in mind. First, specifically ask the vendor what type of wild cards are supported (i.e., prefix, root, and suffix). Many products support only suffix wildcarding. Second, do not become overzealous concerning the need for wild cards. There are other features such as stemming and pattern matching that are more popular among end users because they require less direct input. Many users complain that if they knew where to place a wild card character, they would know how to specify their query term.

Soundex

Soundex is a recall-enhancing tool that executes a phonetic-based search for all user-specified query terms. Thus the burden of knowing an exact spelling or all possible spelling permutations is removed from the user. For example:

FIND RESEARCH.DOC = SMITH

The translation is: "Find all the documents within the text field called RESEARCH.DOC that contain the term SMITH." But, in a system that supports Soundex, any terms that sound like SMITH (i.e., SMYTH, SMYTHE, SMITHE, etc.) will also be searched on. There is no standard for the execution of a Soundex option. Thus, results are somewhat unpredictable. Users typically find this feature too arbitrary to be reliable. The cost in terms of lost precision is often considered too great.

Pattern Matching

Originally this recall-enhancing feature was envisioned as a way to enhance the ability of the inverted index approach to manage spelling errors and differences in an inverted indexing system. In this regard, pattern matching is very similar to the spell-checking functions popular in many word processing systems. If the query term cannot be located in the inverted word index, the system responds with alternative query terms. Selection of the alternatives is based on words in the index that have the most number of contiguous characters in the query term, and the fewest exceptions. For example:

FIND RESEARCH.DOC = CHRYSANTHAMUM

The translation is: "Find all the documents within the text field called RESEARCH.DOC that contain the term CHRYSANTHAMUM." Because the query term is spelled incorrectly, the query execution should result in no documents. However, in a system that supports pattern matching, the user would be prompted to possibly search on CHRYSANTHEMUM. Thus the pertinent documents would be retrieved despite the spelling error.

Pattern matching has been expanded in many products to offer a recall-enhancing functionality similar to wildcarding; however, this approach to pattern matching is more rule-based than wildcarding. Patterns of characters and numbers can be taught to the system, which enable it to retrieve concepts based on these patterns. For example, if the system was instructed to recognize any three numbers followed by a hyphen, followed by any two numbers, followed by a hyphen, followed by any four numbers, as a *social security number,* then users could query for all documents in which a social security number contains the number *4.* Likewise, if patterns which denote names of organizations are taught to the system (i.e., any string of characters with an initial capitalization and ending in *Corp., Corporation, Inc., LTD,* etc.), then users could query for a list of all organizations named in the document collection. In this case, documents would not be retrieved, but a listing of those strings of characters within the documents that represent organization names would be retrieved. This is a popular tool in situations where a user is not clear as to which values to query on (i.e., "What are the valid corporation names in this database?").

Stemming. *Stemming* is a recall-enhancing feature that does not require user input at either the back or the front end. It is completely automatic in those systems that support it, and can dynamically be turned off. Stemming eliminates the need for the user to construct wild card searches or to determine all possible permutations of a query term.

Systems that support stemming treat each query term as a root and automatically search for all logical suffix extensions of the query term as well. For example:

FIND RESEARCH.DOC = RUN

The translation is: "Find all the documents within the text field called RESEARCH.DOC that contain the term RUN." However, in a system that supports stemming, documents that contain the terms RUNNING, RUNS, and RUNNER would also be retrieved. The stemming process is based on standard syntactical rules. Thus, in the example above, the system would also search for the word RUNNED, believing this to be the past tense of RUN. Exceptions to the standard rules of stemming are supported by another recall-enhancing tool, normalization. It should be kept in mind that with few exceptions, stemming algorithms are based on the English language exclusively.

Normalization. Normalization is a higher form of stemming. It also eliminates the need for user-specified wildcarding by automatically tracking all forms of a search term, including those that go beyond suffixing. For example:

FIND DOCUMENT = RUN

The translation is: "Find all the documents within the text field called RESEARCH.DOC that contain the term RUN." However, in a system that supports normalization, documents that contain the terms that are suffix extensions (stemming) and any other forms of the term (i.e., RAN) would also be retrieved.

Heuristic Associations

Heuristic associations are a statistical approach to providing dynamic recall-enhancing capabilities to a text retrieval system. The premise behind this recall-enhancing feature is that there is a body of knowledge regarding word associations and word meanings that can be gleaned from the body of text itself. Thus, heuristic association, unlike the majority of concept-based querying tools, does not require user input in order to build a knowledge base, or intelligence regarding written language.

Heuristic associations however, cannot be utilized on an initial query. This is an interactive tool. An initial query must be supplied by a user. The results of that query can be extrapolated, however, providing insight into the means by which the user's query can be enhanced or expanded to provide a more exhaustive definition of the subject matter that appears to be the essence of the user's interest. For example:

FIND DOCUMENT = BASEBALL

The translation is: "Find all the documents within the text field called RESEARCH.DOC that contain the term BASEBALL." The documents that meet this criterion would be retrieved. In a system that provides heuristic associations, however, a listing of related terms and phrases would be supplied in addition to the returned document set.

Following the same example, the user would be prompted with a list similar to this:

HOME RUN

WORLD SERIES

MAJOR LEAGUE

NATIONAL LEAGUE

AMERICAN LEAGUE

RBI

In such a system, the user would have the ability to select any or all of these suggested terms as the basis for a subsequent search, thus broadening the search or enhancing recall. The list of words that is returned is determined by a statistical algorithm executed against the returned document set, that identifies words and phrases that appear with regular occurrences within the context of the query term. There is no standard approach to this feature. Thus, although several products support heuristic associations, results will vary from one product to another on the same set of documents. It is important that you become familiar with each vendor's approach and methodology as part of investigating a product's offering. In a poorly executed model, users can become frustrated with exhaustively long and predominantly irrelevant lists of associated terms and reject a system that offers little or no insights.

Synonym File. A synonym file enhances the recall of a system. Synonym files, however, are user defined. A product that supports synonym files typically provides an empty file structure capable of holding synonym tracking information which must be entered into the file. These files track groups of words that have like meanings. For example, an entry in a synonym file would resemble:

MAIN TERM: AUTOMOBILE
SYNONYM: CAR
SYNONYM: MOTOR VEHICLE
SYNONYM: HOT ROD
SYNONYM: STATION WAGON

In this example, a query for the word AUTOMOBILE would automatically be extended to include retrieval of documents which contain the terms CAR, MOTOR VEHICLE, HOT ROD, and STATION WAGON. In most systems there is syntax or interactive dialogue which prompts the user to determine whether an exact word or synonym query should be executed. Some vendors provide prebuilt synonym files based on established input, such as *Roget's Thesaurus.*

Synonym files do not track word relationships beyond this single-dimensional model. There is no deductive reasoning used. Therefore, using the same example, the word CAR is not tracked as a synonym of the word MOTOR VEHICLE. It is only the main term, in this example AUTOMOBILE, which has synonyms associated with it. The establishment of cross-referencing synonyms in a synonym file requires that the user create a main term listing for each of the synonyms of every main term.

It is important to note that many products which tout a thesaurus simply supply a synonym file. Much of the confusion concerning synonym file versus thesaurus stems from the fact that in the paper-based publishing world a thesaurus (e.g., *Roget's Thesaurus*) is a synonym facility in text retrieval terminology. Evaluators of text retrieval systems who perceive the need for either a synonym file or a thesaurus, as defined below, should investigate the capability of every product's "thesaurus" before assuming it is indeed a thesaurus and not a synonym tracking system.

Thesaurus. The thesaurus is one of the few areas governed by a standard—ANSI standard Z39.19.[5] Thus, a simple and quick means for determining if a product provides synonym tracking or true thesaurus capabilities is to ask if it is an ANSI-standard thesaurus. Similar to the synonym file, thesauri are not provided by the vendor. Rather a thesaurus file is provided, into which the user can enter terms in a logical fashion such that various relationships between words and word mean-

[5]For a copy of the ANSI standard, "Guidelines for Thesaurus Structure, Construction, and Use," ANSI Z39.19, contact American National Standards Institute, 1430 Broadway, New York, N.Y. 10018.

ings are tracked by the system. A well-conceived and well-constructed thesaurus can enhance both precision and recall.

A thesaurus is a compilation of words and phrases tracking synonymous, hierarchical, and other relationships and dependencies. The thesaurus is based on a series of operators, the most powerful and useful among these being USE, USE FOR, BROADER TERM, NARROWER TERM, EQUALS, RELATED TERM, and SCOPE NOTE. There is no limit to the number of operators used for each term (i.e., a term can have any number of BROADER TERMS associated with it). Likewise, usage of any or all of these operators is not mandatory.

The operators USE and USE FOR track synonyms. The USE operator is used to denote the main term, sometimes referred to as the LEAD or PREFERRED term in a list of synonyms. It is analogous to the main term in a synonym file. USE FOR denotes the terms and phrases that have a meaning similar to the USE term. Unlike synonym files, deductive reasoning is exercised in a thesaurus file. The relationship between the USE FOR term and all its USE terms is implied, so that each word in the grouping is treated as a synonym. Thus whether the user's query contains the LEAD TERM or one of its synonyms, all of the synonyms of the LEAD TERM will be used in the execution of the query. For this reason, unlike in a synonym file, it is not important which term is used as the LEAD or USE term unless the thesaurus will be used to control input as well. A thesaurus has the ability to physically substitute the USE term for all occurrences of its USE FOR terms on input. Typically, this functionality can be turned off. If not being used, selection of one synonym as the LEAD term is not crucially important. (See Fig. 5-11 for an example.)

BROADER TERM and NARROWER TERM operators work in a hierarchical fashion, tracking words that are not synonymous to the lead term, but place it in a broader or narrower definition. Typically, if a query contains a search term which has either or both broader and narrower term(s) associated with it, the user will receive a prompt asking if the broad or narrow definition of the word should be applied. Broader definitions will typically increase recall while narrower definitions will help to increase precision. (See Fig. 5-11 for an example.)

The EQUALS operator potentially increases precision by tracking ambiguous terms, terms that have multiple definitions. If a user specifies a query term that has EQUALS terms associated with it, the user will be prompted to specify which definition of the word to apply to the query. In this way, a sublist of synonyms that are grouped together for this particular meaning of the term is applied to the query execution. (See Fig. 5-11 for an example.)

BROADER TERM: TRANSPORTATION
USE: LOCOMOTIVE
NARROWER TERM: PULLMAN CAR
USE FOR: RAILROAD
USE FOR: TRAIN
  EQUALS: LOCOMOTIVE
  EQUALS: INSTRUCT
RELATED TERM: STEAM ENGINES
RELATED TERM: IRON HORSE
SCOPE NOTES: FOR DOCUMENTS CIRCA 1840 ALSO TRY IRON HORSE

Figure 5-11. A sample ANSI standard thesaurus listing. In this example, the lead term (USE operator) is LOCOMOTIVE. The availability of a BROADER TERM (TRANSPORTATION) and NARROWER TERM (PULLMAN CAR) would allow the user to specify that these terms be used to increase recall or precision respectively. A query which included the term LOCOMOTIVE would automatically be extended using the boolean OR operator to include the USE FOR terms RAILROAD and TRAIN. A query containing either of these terms would respond in kind. If the query contained the term TRAIN, the user would be instructed to choose a definition, specified through the EQUALS operator. In this example, LOCOMOTIVE or INSTRUCT. In this case, the term TRAIN would be used in the execution of the query, but the associated synonyms to be used would be determined by the EQUALS terms chosen by the user. The SCOPE NOTE would be displayed whenever a query contained the LEAD TERM, LOCOMOTIVE. The RELATED TERMS are entered into the thesaurus as a record-keeping, librarian feature, and would only be supplied to the user if they were incorporated into a SCOPE NOTE, as is the case with IRON HORSE in this example.

The RELATED TERM operator is used as a method to increase recall, but not as actively as the BROADER TERM and USE FOR operators. Through this operator, terms that have some relationship to the lead term, but are not synonyms of, broader definitions of, or narrower definitions of the term, are linked to the term and offered to users as other avenues of interest that could be pursued. RELATED TERMS are not automatically executed as part of a query, unlike the other operators within a thesaurus. This is a passive approach to recall enhancement. (See Fig. 5-11 for an example.)

The SCOPE NOTE operator allows information about a term to be stored with that phrase and displayed whenever that term is used in a query. Scope notes can provide information regarding term qualities such as the availability of related terms or dated definitions of a term. (See Fig. 5-11 for an example.)

A thesaurus also must include the ability to be displayed. Minimally, the display should take the form of an alphabetical listing of all lead terms followed by each of its cross-references or links. A benefit of using a standard structure for creating thesauri is their interoperability. A thesaurus built in one product could be migrated to another, if both support the ANSI standard. Third-party vendors have exploited this fact and offer prebuilt standard thesauri, which can be plugged and played with any text retrieval system that supports this standard.[6]

Topic Definition

Topic definitions are very similar in function to a thesaurus facility.[7] There are some innate differences, however, between a topic definition and a thesaurus file. Topic definitions are not based on an industry standard, but are a proprietary approach unique to Verity Corporation's TOPIC system. Topic definitions track term meanings in a hierarchical fashion similar to a thesaurus, but do not support the functionality of related terms and scope notes. On the other hand, topic definitions offer other features and functionality that are not supported by a thesaurus. The structure of a topic definition and the means by which it is constructed and viewed are also different than a thesaurus. Topics are built and viewed using a graphical user interface which depicts the hierarchical relationships between related terms (see Fig. 5-12).

In their simplest definition, topics, like a thesaurus entry, are words or phrases that are related in a hierarchical manner. However, the topic definition is not limited to terms and phrases. Each node of the topic can be any instruction understood and executable by the search engine. Thus, a node can specify a boolean expression, a date range, the use of stemming, wildcarding, a soundex search, etc. Topics can be nested inside other topics to provide a true hierarchical approach.

Additionally, each node of a topic tree can be assigned a weighting factor, a system which is used in relevancy ranking. (For more information on relevancy ranking and term weighting see the section on relevancy ranking in this chapter.) It is important to remember that topic definitions are not preconfigured. A file capable of holding these definitions and front ends facilitating their creation are supplied. The user must define the topics and enter them manually.

[6]Consult *Thesauri Used in OnLine Databases: An Analytical Guide* by Chan and Pollard, Greenwood Press, 1988, for a listing of commercially available thesauri.

[7]Although several vendors use methods similar to this, the term TOPIC is a registered trademark of Verity and topic definitions are proprietary to the TOPIC text retrieval system from Verity, Inc.

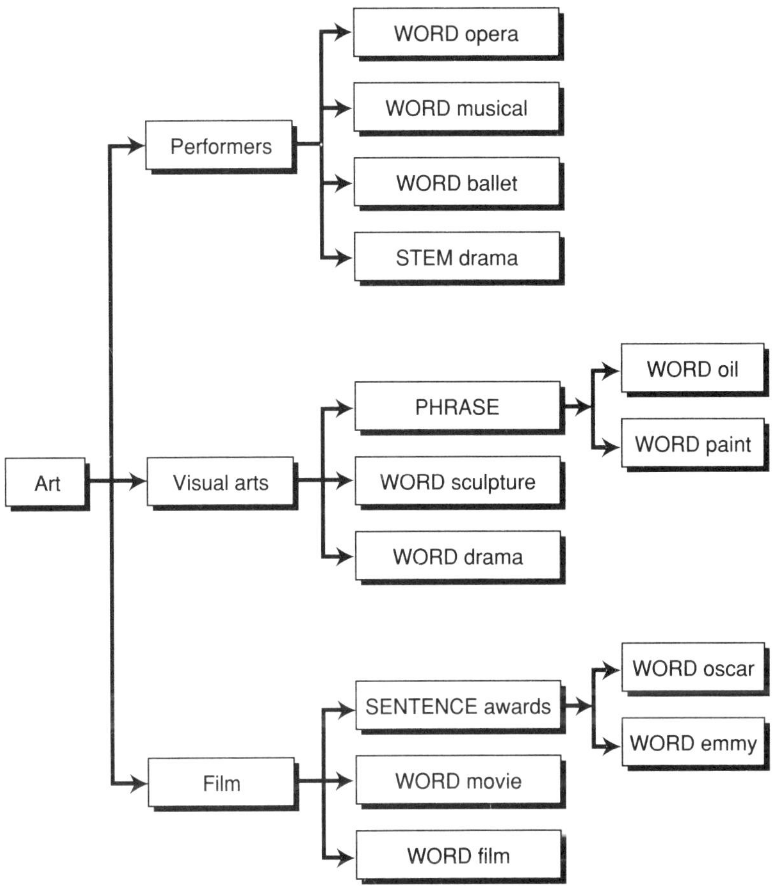

Figure 5-12. A sample of topic definitions. Topic definitions are similar in function to a thesaurus facility. Topic definitions are rule-based objects that represent areas of research or interest. These objects consist of a hierarchical arrangement of words, phrases, and search criteria related to the topic. Each node of the hierarchy is given a weight representing its overall importance to the topic. Queries can be constructed to retrieve documents based on actual words via an index, or using these defined topics. By querying via the preexisting search objects, users can take advantage of expert knowledge and insight.

Semantic Networks

The semantic network is an approach that automatically tracks the relationship between words and phrases. In many regards it offers functionality similar to the topic tree and thesaurus. Perhaps one of the biggest differences, however, is the fact that the semantic network is integrated into the search product—prebuilt and available out of the box. The intelligence contained in a semantic network is supplied by the vendor. The end user is not required to impart any knowledge into the system, but if the user has a need to modify or enhance the semantic network, tools are available to make this possible. For example, semantic networks will accept an ANSI thesaurus file as input which would alter the overall knowledge and structure of the network.

Semantic networks are not that dissimilar from neural networks but they rely more on traditional ideas of semantics. A myriad of rules goes into a semantic network, but the principal rule is that you are evaluating the sentences, structures, and grammar in the documents, not just the words. In a semantic network word meanings are tracked using a form of set theory mathematics. Relationships (synonyms) between words are tracked. More powerfully, definitions which arise from the concurrence of words are also tracked. For example, STATE has several definitions associated with it, as does ART. But, the expression STATE OF THE ART is also recognized as a term which has a meaning different from the sum of its components.

The availability of this level of knowledge makes it possible for a user to ask naive questions and get back sophisticated, concept-based results. Thus semantic networks control both precision and recall in a text retrieval system. An additional feature is the innate support for a natural language query. Because the semantic network attempts to understand the document text as a whole, it requires nothing more for a query than a string of text.

I have found that the best way to describe the capabilities of a semantic network is to share with you my first encounter with such a system. I was given the King James' version of the *Bible* in a semantic network-enabled text retrieval system. A natural skeptic, I approached the product with great cynicism. I decided to avoid obvious questions such as: Who was Noah? or Who was turned into a pillar of salt? I posed what I believed to be a question without answer: What does God do for a living? (It is interesting to note that this is exactly how I typed in the query.) Within seconds the system came back with several sections from *Genesis,* highlighting passages such as:

> "On the third day He separated light from darkness."
>
> "So God made the sky, dividing the vapor above from the water below."
>
> "So God created great sea animals, and every sort of fish and every kind of bird."
>
> "So God made man…."

Needless to say I was amazed. I had never looked at it from this particular angle before. God did indeed have a job! The semantic network pointed this out to me through its objective unbiased analysis of text and query. Further investigation uncovered that the semantic network understood the phrase *does for a living* to have a meaning similar to words and phrases such as *makes, creates,* etc.; thus the connection to the passages in *Genesis.*

A great example of the flexibility and potential power of the semantic network is the information refinery. (For more detail on the information refinery see the section on related features in this chapter.) In an information refinery what you are looking for is semantics in a document. For example, I am looking for evidence of valuation of a company that suggests a potential takeover target. In the document that might be expressed by the term "the stock of corporation ABC has gone up in the recent past by more than 50 percent making it a prime target for potential suitors." The document doesn't explicitly state "takeover" but it speaks to that. If I take apart the semantics of that sentence, I could probably infer that something positive is happening to a corporation, that a positive impact is being expressed against the stock of the corporation which is typically the way I value an organization, and as a result a high value over a short period of time could result in a potential takeover. The semantic network makes similar evaluations and thus would return this example document in response to the sample query.

Similar to the concept-based clustering search algorithm to text retrieval, perhaps the biggest end-user benefit from semantic networks is that they innately allow the end user to think of his or her query in a free-thinking, unrestricted manner. Experimentation and probing is encouraged. Unlike the clustering approach, the semantic network makes this possible without requiring that the user input knowledge into the system initially. The greatest shortcoming to this text retrieval feature is the fact that it is developed only for the English language.

Relevancy Ranking

In the game of golf the object is not to hit the ball but to hit it with just the right amount of force and with just the right club. Occasional golfers may have the form of Arnold Palmer but all too often they try to hit the ball with the strength of Arnold Schwarzenegger. The trees that line the fairways are littered with golf balls—a testimonial to the frustration of the game. The same phenomenon occurs when unwitting users first step up to a text retrieval system and attempt to search several million pages of text with a one- or two-word query. The result is a few hundred documents and a very frustrated user. This is undoubtedly why relevancy ranking consistently ranks higher in necessity among users of text retrieval systems than evaluators of text retrieval systems.

Despite the variety of tools we have discussed that enable text retrieval users to enhance the query process, ultimately the problem with most query results is the amount of information retrieved—usually more information than can be effectively absorbed in the available time frame. This is where the facility to relevancy rank becomes key. In fact, we could go so far as to say that relevancy ranking is the only way to optimize the mix between precision and recall since it will present all relevant information yet also provide the opportunity to prioritize the retrieval of the most relevant information.

Relevancy ranking provides a level of control over precision by presenting retrieved documents in the order of their relevance to the search query. Without support for relevancy ranking, all documents that meet a search query in any capacity are presented in an arbitrary order. Relevancy ranking provides a level of intelligence over the retrieval process. This can become critical in large document collections in which the chances of a query resulting in a large number of documents retrieved are high. A study conducted at the University of Virginia found that our ability to perceive relevancy or value in textual information is diminished if we are faced with more than 15 sources. Your ability to recognize a relevant document is seriously impaired beyond the fifteenth document. Thus the benefit of the retrieval can be lost in a sea of documents which the user has to read or at least scan to determine the net worth of each. A ranked list of documents allows users to conduct their searches in a controlled and intelligent environment. The research begins with those documents that appear to bear a stronger connection to the query. Relevancy thresholds can be set so that a document will not be returned to the user unless its ranking is above a defined value.

There are a variety of ways that relevancy can be assigned. We will review some of the more popular alternatives but none are exclusive of the others. In many cases it is preferable to have a variety of ranking methods so that users can choose the one best suited to their needs. In some cases several can be used concurrently. For instance, you might decide to relevancy rank based on date for your initial query; then to relevancy rank based on word frequency for a subsequent query; and finally, if there are still too many documents to review, to relevancy rank based on term weights. By using different methods of relevancy ranking it is also possible to determine which documents appear to be relevant with each method, and in this way further increase precision and recall by providing multiple perspectives on the documents. Most importantly, however, you should realize that when evaluating text retrieval products, it is not enough to determine if a product provides relevancy ranking. You must determine how relevancy is calculated in the product and choose the approach that best mimics the manner in which your users would determine relevancy.

Date. The oldest and easiest approach to ranking relevancy sorts the retrieved documents by their creation date. The assumption in this approach is the newer the information the more valuable it is to the research. In situations where information's value is closely tied to time and where there is already a high degree of precision demonstrated, this is a valid stand-alone approach. However, where the content of the information itself is equally if not more important than its age, relevancy ranking must go beyond this approach and base decisions on document content as well.

Term Summing. *Term summing* is the simplest and most straightforward content-based approach to relevancy ranking. It is the most widely available, perhaps because of its simplicity. This approach sums the occurrence of every query term in each document. The document with the greatest number of query terms present is ranked as the most relevant. The philosophy in this approach is the greater the number of query terms present, the higher the probability that the document's content is concerned with the query topic.

While this approach has merit it can produce unexpected results, especially in situations where users have difficulty with boolean OR and AND usage. This is also true in databases in which there is great discrepancy in document lengths. Longer documents have a high probability of high query term counts. But this does not necessarily mean that they are more relevant than a shorter document with proportionally more query terms.

Term summing also does not take into consideration the fact that information can be greatly dispersed in large documents. For example, searching for the terms SPACE and EXPLORATION may result in many document hits in which the terms are separated by several paragraphs or pages of text, thus not necessarily relevant to the query. For these reasons, algorithms which take into consideration term proximity and density are often necessary. (An explanation of term density and term proximity can be found later in this chapter.)

Weighted Term Summing. Building on the approach of term summing, and potentially eliminating some of its weaknesses, are algorithms that use and permit specific values known as weights to be associated with query terms. Thus, the totaling is not done on a word occurrence basis only, but also on the weights of word occurrences. Simple term summing adds a constant value (e.g., 1) for each occurrence of any query term found in a document. In a weighted-term-summing approach a term's weight is added to the document's relevancy accumulator for each of its occurrences.

The weight value assigned to unique occurrences of words will vary among implementations of this algorithm. Some products permit term weights to be dynamically altered by users as the text retrieval system is used. This results in a user-specific view of the relevancy in the document. These tools may also allow for multiple user weights and thereby individualized views of relevancy. This provides different perspectives on the same set of documents, which can prove to be very insightful, especially in knowledge-based applications where expert experience is a key component of effective research.

Term weighting provides a means for identifying the importance that specific words have to an overall database, a search query, or a topic. Though several approaches are used to store, scale, manage, and use term weights, basically they are used to provide insight on how terms affect the value and uniqueness of retrieved documents. This lexical control is used in conjunction with relevancy ranking algorithms to provide greater precision during document output.

In one approach, term weights are assigned automatically to every word in the inverted index (thus this approach can only be used by systems that use this search methodology). The weights are based on a scale (i.e., 1 = not important, 100 = very important). The weights are assigned to words based on the word's frequency in the database; the more prevalent a word is in the database the less important it is perceived to be. The philosophy here is that infrequently used words convey a uniqueness to the documents they are contained in and therefore provide a means to differentiate this information from the mass.

Another approach works in generally the same way, except that term weights are not assigned to the terms until a document set has been identified by a query. The philosophy here is that words found infrequently in a set of documents convey a uniqueness or importance in this set of documents.

The most popular approach to term weights allows the user to assign the weights as part of the query. By allowing users to assign term weights to query terms, similar type results are achieved, but the user can impart human knowledge concerning the importance of terms rather than relying on a statistical analysis. For some users this is a critical difference because it is their knowledge that they can impart on the logic of the ranking algorithm. This feature can be used in creative ways to minimize network impact and the time spent reading through documents. For example, a user could execute a query using a particular set of term weights. Without reading any of the documents the user could submit the same query but change the term weights. The resulting set of documents will be the same but they would be ranked differently. After several such exercises, the user's insight into the content of the documents is very keen, allowing the user to make strategic and calculated decisions regarding which documents need to be opened and read.

Topic trees provide another dimension to term weights. Term weights can be assigned to the individual nodes of a topic definition. In this way, term weights are not linked to the entire database or to a set of query terms, but within the context of a subject matter. Terms can appear in several topic definitions and can be assigned different weights in each. This approach to term weighting supports the concept of weight accrual. Accrual allows the combined weights of several linked terms to be treated as a single weighting factor, putting this multiterm concept on equal footing with individual terms. The concept-based clustering search methodology also supports term weighting as part of its structure. In this approach terms are weighted against subject matter categories, reflecting the level to which the word connotes the subject.

Omni Term Skewing. While the basic premise used in this approach is the same as that of term summing, a skewing factor is used to recognize the importance of those documents that contain at least one occurrence of each query term. For example, document A has 12 total occurrences of query terms, while document B has eight total occurrences of query terms. Simple term summing would rank A more relevant than B. But the user's query was based on three separate query terms. Among the 12 occurrences in document A, only two of these terms are found. Document B contains at least one occurrence of each of the three query terms among its eight total occurrences. In this case a skewing factor would be added to document B's total, representing the unique value of having all terms present. This is especially helpful in situations where users will precariously link together several terms in an effort to increase recall.

This approach increases in its effectiveness, as do all term summing algorithms, as the number of query terms increases.

Fuzzy Logic. The premise behind omni term skewing is enhanced through the application of *fuzzy logic* (not be confused with fuzzy searching which is a feature of the pattern recognition search methodology). Fuzzy logic in a text retrieval system is a relaxation of the boolean AND operator. For this reason, fuzzy logic is sometimes referred to as "soft" boolean logic. The approach is especially helpful in situations where the boolean OR and AND can be misused, or perhaps misapplied by novice or casual users. Fuzzy logic was originally introduced in reaction to the starch criticism behind using boolean logic to perform content-based queries. Many industry experts argue that boolean logic is far too rigid to be effectively applied to text retrieval.

For example, consider that you are interested in doing research into the effects of nuclear radiation on eyesight. You would logically construct a query similar to:

FIND DOCUMENT = NUCLEAR RADIATION AND EYESIGHT

The translation is: "Find all the documents within the text field called RESEARCH.DOC that contain the term NUCLEAR RADIATION and the term EYESIGHT." The documents that meet this criterion would be retrieved.

In this case, documents that address only NUCLEAR RADIATION or only EYESIGHT would not be retrieved. Yet, many argue that the fact that a document bears some light on the subject makes it potentially valuable. Therefore, fuzzy logic makes the document available.

Fuzzy logic figuratively reduces all boolean AND operators into boolean OR operators. Thus documents are retrieved if either of the query terms are present, despite the user's request for both terms. The user's perspective (need for both terms) is applied as part of the ranking, however. All documents that meet the AND condition are automatically ranked higher than those that do not, despite any other summing algorithms that may be applied.

Another approach to fuzzy searching can be found in systems that support stemming, normalization, synonym tracking, and thesaurus. The occurrence of the exact search term in a document is given a higher value than the occurrence of one of its forms or tenses, synonyms, or broader or narrower terms. Assuming that the users of the system carefully choose the terms used in a query, the application of this rule in the relevancy algorithm is valid. However, in situations where stemming, normalization, synonym tracking, and thesaurus are used expressly because users do not always specify all the possible search terms in their most prevalent form or tense, this application has little merit.

Term Density. In each of the term summing approaches there is no mechanism for taking into account the density of query terms in a document. For example, a document of 35,000 words may contain 100 occurrences of query terms and a document of 2,000 words may have 40 occurrences of query terms. Using term summing algorithms, the former document would be considered more relevant by nature of its 100 contained terms. Yet there is a possibility that the latter document is more relevant to the query since its query term density (percentage of total document composed of query terms) is higher. Therefore algorithms that support term density will compute this value and use it to further define the relevancy rating. As mentioned in the discussion on term weighting this is a critical feature in databases where there is a broad range in document size.

Term Proximity. Term proximity is similar to the logic of term density. Term proximity adds additional value to the document for every occurrence of query terms positioned "close" to one another. Closeness is typically defined in terms of number of terms apart. The exact definition and value assigned will vary among products that support this approach.

One of the principal benefits of term proximity is its effectiveness in small queries with few words. Many studies of text retrieval users have shown that in almost all cases novices and casual users will use two or three terms in their query. They don't use extensive boolean logic because they don't want to be bothered with having to think through the query logic—they just want an answer. Because the query will result in an overwhelming amount of information due to the probabilistic effect of word frequency in large documents, proximity is a requirement for basic text retrieval on large document databases.

Concept-Based Clustering. The concept-based cluster search methodology discussed earlier in this chapter innately supports relevancy ranking. Concept-based clustering system implementation requires the user to manually rank the relevancy of terms to subjects. Using this information a virtual storage area is constructed into which documents are positioned, representing their overall subject matter. User queries are subsequently processed and positioned into the storage area in the same manner. Document relevancy is determined by the proximity of the document to the query in the clustered storage space. The nearer the document the more relevant it is assumed to be. (See the discussion on concept-based clustering in this chapter for more details.)

Link and Node Weighting

Although technically not a form of relevancy ranking, this discussion would not be complete without mention of link and node weighting. These are features of a hypertext system. Similar to the concept of term weighting in text retrieval, link and node weighting will enable users to ask questions of the system such as "What is the best path for me to pursue?" Node and link weighting supports the assignment of numerical values to links and nodes. These values are rated on a scale indicating relevancy and importance. Node and link weights can be determined automatically by the system based on usage or an internal algorithm, or manually assigned.

Weighting will be the foundation for other navigational aids. One of these is browser thresholding. Browser thresholding takes advantage of the node and link weights and uses them to answer inquiries like "What are the most relevant paths to my research?" or "Show me just the very important information." The browser facility could respond with a graphical depiction of just those nodes that are weighted high enough to meet the inquirers' desired level. This could eliminate guesswork and hours of text scanning on the part of the end user. Currently, however, this functionality exists only in the design lab.

Other Text Retrieval Features and Issues

The underlying search methodology and query-enhancing tools are the foundation of a text retrieval system. Initial product scrutiny must be focused on these facets of the product. But text retrieval environments typically have to provide functionality and features beyond document retrieval itself. Although examination of search methodology and query tools will provide you with a short list of vendors, it is the additional features that typically will help you decide which product to choose. Additional features needed must be determined through a careful analysis of the application and/or business model. What will your users do with their documents once retrieved? From where are the documents coming? How much storage overhead can your system support? These issues are addressed through additional product features which will range from compression algorithms used on the document text to support for text editors.

Output

Virtually no text retrieval system displays documents in response to a query. Rather, statistics regarding the number of documents found that meet the search criteria and the number of query terms found in those documents are displayed. This is a benefit, especially in cases where access to the document collection is over a network. Impact on the network is minimized by displaying these statistics as opposed to transporting all the documents retrieved to the user's workstation. Very often, a listing of document titles or some other such information will also be displayed. In cases where relevancy ranking is provided, this feature is very useful. Based on perceived relevancy and document titles, users can intelligently select just those documents that they want to view.

This type of information also provides support for precision and recall control by giving insight into the document set before it is viewed. Based on this information, users may decide that a broader-based or a narrower-based query should be used. The more information provided by the system the more decisions of this type can be made by the user prior to viewing documents.

The manner in which documents are displayed will vary as well. As a minimum, every system will display the document in its entirety. Some products will also provide the ability to display just those portions of text in which the query terms were found (the portion size can be user-controlled). Generally, query terms are highlighted and a function key provides the ability to jump through the document from one query term to the next. When perusing a lengthy document this feature can be critical. Imagine plowing through a 50-page document just to find that the portion of text that caused the document to be retrieved was on the last page.

Some products will display the document in its original format using its native editor as the display mechanism. Be aware that other products provide this capability, but when used, the ability to highlight query terms is lost.

In databases containing compound documents, you must determine the infrastructure that supports the linking of images and other non-text forms of data to the text file. Are images displayed embedded in the text, similar to the manner in which the document was produced, or are they linked to text using some form of hypertext technology? The latter is the predominant approach used today, but industry trends are toward the former. Growing support for and advancements in standard compound document architectures should accelerate the ability to perform search, retrieval, and display of true form compound documents. (For more information on these standards see the related standards section of this chapter.)

In cases where images and text are linked, you may need to determine if it is necessary to initially view the text in order to obtain any related images. I recall one client who needed to query brochures based on the content of description within, but wanted to view the pictures (images) contained in those documents. Products which forced the user to view the text in order to get to the images were considered too cumbersome and thus eliminated from consideration.

The recent emergence of document viewers is having an impact on this facet of text retrieval. The document viewer stores the document in a proprietary file, but one which can be displayed on virtually any platform. The viewer also preserves the original format and style of the document. Early criticism of these viewers revolved around the size of the files, the lack of sophisticated retrieval capabilities, and the level of granularity they impose. (Viewers typically treat each page of a document as the smallest unit; thus, linking and viewing are forced at the page level.) As these viewers mature, they are meeting these shortcomings head-on, and overcoming them with the application of various text retrieval technologies and document-handling architectures. It is reasonable to expect the document viewer to redefine the competitive arena of document output.

Support for hard-copy output (printing) offers a myriad of choices as well. Functionality runs the gambit, including: print screen capabilities only, print the document in its entirety, print just those sections of the document which are pertinent to the retrieval (sections can be automatically or manually defined), and print in a variety of output styles made possible by the availability of an integrated report writer. In cases involving compound documents, you should ascertain whether the document is printed in its true compound structure, or if linked graphics and images are printed on separate sheets of paper.

Document Modification

In most systems, retrieved documents can be edited through either a proprietary editor or the editor native to the document. The latter approach is the popular trend in the industry, the exception being those products geared towards hypertext on-line publishing, in which proprietary editors are prevalent.

In systems that employ an index as a means of retrieval, you must concern yourself with the effect that document modification will have on the index. Retrieval accuracy will only be as timely as the index. Thus, an index updating strategy must be in place. Index updates can be accomplished dynamically on-line or through batch mode. In a hypertext environment, modifications to the documents might also require that the links be maintained. Systems that are dynamic and require frequent updating should be developed in text retrieval systems that minimize this overhead. (For a detailed discussion regarding index maintenance and its effect on retrieval accuracy, refer to the section on search methodologies in this chapter.)

Electronic Sticky Notes

A form of modification that does not physically affect the document, electronic sticky notes capability is quickly becoming a "must have" in most text retrieval applications. Utilizing a form of hypertext technology, this feature allows you to attach notes of text (or sound in some products) to a document, in much the same way that stick-on notes are used with paper documents. The note is represented as a hypertext button. Enacting the button makes the note visible with the text of the actual document but is not considered a part of the document itself. The ability to index and search on note text is product-dependent. In more sophisticated approaches, security can be placed on notes, separate from the security on a document. In this way, you can attach a routing slip or comment meant for certain individuals only, despite the general availability of the document to which it is attached.

Intelligent and Live Documents and Information Agents

Over the last several years, through the integration of AI techniques, text analysis, and sophisticated linking structures, the document has taken on many new properties and features. Though technically not all of these are directly text retrieval features, no discussion on text retrieval and document management would be complete without their mention.

An all-encompassing term used to refer to this new document form is the *intelligent document.* The intelligent document is a dynamic information resource. This is the most basic difference between the intelligent document and the paper document or the electronic document which is nothing more than an on-line version of a printed page. The intelligent document is not a repository of static information, but a container of pointers to various sources of information that are subject to modification. Thus the document matures as its information assets do. In this way, the document becomes an evolving and intelligent entity capable of keeping itself up-to-date. In the text retrieval arena, the intelligent document is further refined and enabled along two separate but not mutually exclusive applications of technology: linking and information agents.

Linking, the mechanism used by the intelligent document to track and monitor the components of the document, takes two forms: *live links* and *intelligent links.* Live links are links to a relative file and its application. By utilizing live links, the document automatically retrieves and displays the latest revision of each of its components. Many hypertext systems are supporting this form of linking. The live link can transcend levels of processing. For example, if a document contains a bar chart which was created by a spreadsheet package and a change is made to the spreadsheet data, a new version of the bar chart would automatically be generated and placed in the document. With intelligent links, the level of knowledge imparted on the document is much higher. Intelligent links not only track the latest version of the document's components, but the relationships between components. A change in one component may cause an automatic change in another. For example, a bar chart is complemented with a caption depicting its contents. A change to the caption automatically updates the chart or vice versa. At the very least, these links will issue a warning when the document is accessed, indicating that components are potentially out of sync.

Information agents (also known as intelligent agents or knowledge agents) utilize sophisticated text retrieval approaches to make the document self-aware. This is particularly useful in databases that utilize live documents, contain documents that are subject to frequent change, or situations in which the database is continuously added to through integration with a live-wire feed (i.e., UPI). User profiles, which take the form of a text query, run in background on an ongoing basis, triggered by the update to a document or the advent of a new document. Based on the users' needs, the information agent dynamically determines which documents need to be routed to which individuals in a real-time manner. With information agents, the onus of communication is shifted from the authors or managers of information to the readers and consumers of the information. The key to information agents is the underlying search methodology and technology approach to determining the basic content and value of information within a document. Several approaches exist, ranging from simple inverted indexes to sophisticated text analysis tools.

An advanced form of the information agent is the information refinery, also known as staff technology and text abstracting. Information refineries not only ascertain the value of information to particular individuals, but also interpret the data and present it in the most logical or expedient manner. For example, if a document contains extensive text regarding sales trends, the refinery may reduce the information to a single pie chart and route the chart, hypertext-linked to the original document, to interested parties. A 30-page document may get abstracted into a three-paragraph overview. The refinery can also be used to spot trends across incoming documents and use this information to make calculated predictions, or suggest alternative and related areas of interest to a user. In other cases, two subjects can be presented, and the refinery asked to determine if there is a connection between them. For example, a law enforcement official, using a document database of police reports, crime reports, and the like, could submit a query in the form of JOHN DOE AND ILLEGAL ELECTRONIC FILE TRANSFER, IS THERE ANY CONNECTION? Using various statistical, text analysis, and hypertext algorithms, the information agent could find a connection (if one indeed exists) buried within the volumes of text, and present the evidence to the user as a single virtual document. (For other approaches to information refineries see the section on semantic networks in this chapter.)

Compression

As EDMS technologies are paving the way for acceptance of on-line documents, the matter of overhead is becoming a growing concern. Specific to the arena of text retrieval systems, the greatest concern is in the area of index overhead. In response, vendors have introduced compression algorithms which minimize index size. In some cases the compression algorithms used on the word indexes are also applied to the text documents. In these situations, there are reports of entire databases (documents and text) that occupy 30 percent of the size of the documents in their native formats alone. Data and index compression can be beneficial to environments where on-line storage capacity is an issue. If you are concerned with overhead, then your product evaluation should include careful analysis of any compression available in a product. It is not enough to ascertain if the product uses compression. You should determine the compression ratio, whether the compression is applied to data as well as index, and if the compression and expansion routines result in significant processing overhead and delays. There are basically three approaches to compression.

1. The front-key compression algorithm is applied to word indexes in which the words are stored in alphabetical order. Front-key compression, recognizing that in alphabetical listing there is a high probability of prefix redundancy from term to term, eliminates the redundancy and stores only the unique portion of subsequent words along with the length of the common prefix.

For example, a front-key compression index for the words *strategic* and *strategy* would look something like this:

STRATEGIC
<7>y

Using this compression technique alone, indexes can be kept to approximately 100 percent the size of the text in the database. However, this technique can be combined with other techniques to decrease the size of the index even further, while simultaneously decreasing text storage requirements as well.

2. One such technique is the *Dictionary-based Compression System* (DIBACOS). DIBACOS is based on the premise that two-thirds of the words in any document are represented by 1024 of the most commonly used words in the English language. Under DIBACOS, documents are converted to a format wherein these common words are represented by an integer value (e.g., a 10-bit address) in the documents themselves as well as in the index. DIBACOS can typically compress text and indexes at 60 percent.

3. Another compression technique that can be applied to the index as well as the text is Huffman coding. Huffman coding can be performed at a word or character level. The text in the database is surveyed in order to compile statistics regarding word or character usage. A tree structure is built in which the more commonly used a word or character is, the further up the tree structure it is located. Bit values are then assigned to the nodes of the tree by assigning 0 to all left branches and 1 to all right branches. The paths to each individual word character or word become the bit representation in both index and text (or just text in a system that does not use an index). Thus more commonly used characters and words are represented by the shortest bit values. Huffman codes can be constructed as part of the initial data loading process and remain stagnant, or can be continually monitored and updated.

The Future: Dynamic Document Clustering

Despite its longevity, text retrieval technology continues to undergo refinement and radical changes. Semantic networks, now an accepted feature of many systems, were considered theoretical technobabble only a short time ago. There is every reason to believe that new retrieval environments and tools will continue to emerge. An example of this that holds great promise is dynamic document clustering. This is an area of retrieval technology that holds great promise for increasing the recall capabilities of text retrieval. In development for several years, the premise of dynamic document clustering is that knowledge is imparted to the document database heuristically, through the use of documents. The basic architecture of dynamic document clustering is similar to that of the concept-based document clustering model discussed earlier in this chapter, except that dynamic document clustering is not based on predefined rules or expert systems.

In this model, document clusters that represent the relationships between documents are defined by the retrieval dynamics of the system. In such a system, as the textual database is queried through a traditional retrieval engine, the frequency of documents being retrieved together and similarities in the documents (i.e., word frequencies, linguistics, and other patterns) are tracked. These common usage statistics are used to organize the storage of the documents within an *n*-dimensional virtual storage area. Documents that demonstrate a relevancy to one another are stored (logically) closer together. The greater the relevancy factor between two documents, the closer together they are stored. Conceptually, the documents all reside in a multidimensional space. The relevancy factors affecting the placement of documents in this space are dynamically updated through this constant monitoring of user queries and document activity.

The benefits of query results abetted by dynamic document clustering are diverse, but the most salient relate to its ability to bridge the gap between separate but related subject areas and to recombine documents from two distinct areas into a new information base.

This aspect of dynamic document clustering known as recombinant information management can yield dramatic results. It entails joining different pieces of data which would otherwise not be correlated. Since new knowledge is being discovered rather than existing pockets of information being gathered, the end result is often a whole which is far greater than the sum of its parts. In effect dynamic document clustering automates the "think tank" approach to problem solving. This is especially powerful if your document collection is large and diverse. For example, consider a library of documents covering everything from twentieth-century philosophy to electronic document management systems. Though these are separate and distinct subjects, there is the possibility of overlap.

Consider that the foundation of document clustering theory is rooted in the school of philosophy known as poststructuralism. Thorough research on document clustering would not be complete without an understanding of this school of philosophy. Yet most if not all novices to the field of document management will undoubtedly overlook this connection. Queries for information regarding document management and clustering would probably never retrieve articles on poststructuralism. But, if research were conducted in a dynamic clustering environment, expert insight into the field of document management and clustering would undoubtedly manifest itself in several queries that contain terms that link these two subject areas. These queries would provide input to create clusters that would be available to the novice, who when querying on document clustering would receive the insight of the expert on the resulting document set.

In order for dynamic document clustering to become a practical solution, two fundamental problems need to be resolved: cluster maintenance and processing requirements.

The obstacles associated with cluster maintenance center on the issue of controlling the integrity of the clusters. Not every query will necessarily exhibit the expert-like results that should be reflected in the cluster formation. Queries executed by nonexpert users should not be considered part of the system's experience set. A security facility where nonexpert users are assigned to specific groups whose queries are given low-priority weighting, or excluded from the clustering process, is one potential method for handling this situation.

In the case of the expert user, it is foreseeable that through trial and error research, false hits or less than desirable results could be obtained from time to time. It can be argued that since we are looking at expert usage, the occurrence of this is not likely to be frequent enough to dramatically and adversely skew the formation of clusters. In any case, however, it is feasible that a screening procedure could be implemented to allow those users with the proper security rights to decide whether a particular query result set should be applied to the system's intelligence.

The execution time and memory requirements of the algorithms used in document clustering are significant. In a heuristic clustering environment in which the clusters are potentially reevaluated with every query result, this becomes a major resource issue.

Despite these drawbacks however, dynamic document clustering holds tremendous promise. An environment such as the Internet is an ideal breeding ground for dynamic document clustering. It is the intersection of two powerful forces, a large user population with the potential of generating combined expertise through query, and a large collection of seemingly disparate documents. The focus and acceleration in usage of the Internet and America's technology superhighway may provide the incentive needed by vendors and thus fuel research and development in this area of text retrieval technology, ignored for too long.

Associated Standards

Text retrieval is a technology that is predominantly riddled by a lack of standards. In fact, it is interesting to note that where standards exist, in many cases, the standard was driven by another means or is vicarious to the text retrieval process. Examples of these standards are: the ANSI standard thesaurus, Z39.19, which was originally developed as a document classification standard for the library industry; and Huffman coding, whose inception was in the traditional database world. Two other examples, which have not been discussed in this chapter, are ODA and SGML.

Open Document Architecture (ODA) is an ISO standard storage vehicle for compound documents. Though applicable to the text retrieval market, it is more closely linked with the document management arena. ODA promised to create a standard metaformat for all document types (not just text). It was backed by the large system vendors such as IBM and Digital Equipment Corporation. While vendors were prepared to accept documents from the format standard, however, few were providing a means to export to it. The momentum behind ODA suffered as a result. While ODA received a shot in the arm by the formation of a new group, Open Document Architecture Consortium (ODAC), this effort has yet to generate enough momentum or attention to put ODA in the forefront. The functionality addressed by ODA is also being addressed in other approaches such as Apple's OpenDoc and Adobe's Acrobat. Thus the need for ODA per se is waning as well. The point to be made here is that such document standards are peripheral to the text retrieval market. Text products will need to support any document file standards, but the driving force behind these standards typically does not emanate from the text retrieval market.

Such is the case with SGML as well. SGML's roots are in the publishing arena. It provides a coherent and unambiguous syntax for describing a document's structure and formats whatever a user chooses to identify within a document (i.e., rules for paragraphs, titles, subtitles, and quotations). The benefit is the interchangeability of the SGML document. SGML-compliant documents can be accessed and processed on platforms using different word processors and editors without loss of formatting structure. The somewhat sudden propulsion of SGML into the general document management market is much attributed to the U.S. Department of Defense (DOD) because of the CALS initiative. CALS supports SGML as the document standard component of its strategy for storing and sharing textual information. The size and influence of DOD has positioned SGML as the industry de facto standard. Virtually every text retrieval vendor has an SGML handling strategy in place or has announced plans to do so.

Perhaps the strongest evidence of the popularity and applicability of SGML to the EDMS is the fact that the SGML standard has been expanded to include support for hypertext documents. HyTime is an SGML standard that supports the integration of synchronized playback of multimedia elements in a document. HyQ, a component of HyTime, is a proposed language standard for text retrieval query, which is especially architected to utilize SGML structure in a document. (For more detail on HyTime refer to Chap. 6.)

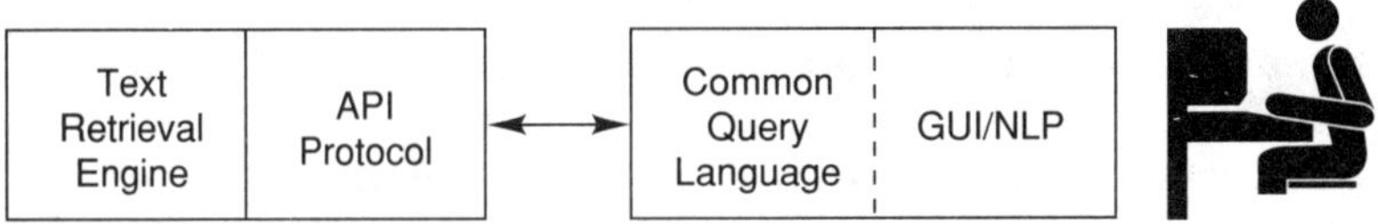

Figure 5-13. Architecture of the text retrieval query environment.

HyQ is not the only proposed text retrieval query language standard. In fact, this is the most active area of text retrieval in regards to standards. All of the activity in this respect has grown out of and is directly linked to the text retrieval community, in direct contrast to the history of the other standards discussed. The imposition of a standard query to text retrieval systems is best understood if you first conceptualize the text retrieval query occurring at three levels: the GUI level, the query language level, and the API protocol (or processing) level. (See Fig. 5-13.) There is standard-related activity happening at each of these levels.

The GUI level is the front end with which the user interacts. Although a command-level language can be positioned in this area, a popular trend is to front-end any query language (proprietary or otherwise) with tools such as natural language query, query by example, or query by forms to accelerate user acceptance and decrease the need for user education. (For more information on these front ends see the text retrieval query section of this chapter.)

The common query language layer can be interfaced by a GUI, but whether or not it is, this layer provides the syntactical command-driven environment by which queries can be submitted. This is separated from the API layer, which, as a layer of functionality, provides an approach to processing the user request, packetizing the resulting information set, and transporting it across a network. While you need not become too concerned over the architecture and relationship between these layers, they are mentioned only because there are standards proposed for each layer. To properly understand and make decisions regarding your organization's direction regarding text retrieval standards, knowledge of the layers is required. End users, however, need not become concerned with the technical background, as the integration between these layers will be automatic and transparent.

At the common query language layer there are three proposed standards: Z39.58, CD-RDx, and SFQL.

Z39.58

Z39.58, better known as the *Common Command Language* or CCL, is a query language proposed by the National Information Standards

Organization (NISO) for use with on-line information retrieval systems.[8] It was developed explicitly as a text retrieval query language. Therefore, in addition to its 19 basic commands, the nature of the language vicariously mandates the support of other features in a full-text retrieval system. Examples of these are:

- Query results must be maintained at least throughout a session.
- Query results can be combined using boolean operators.
- The result of a query is the number of records retrieved, not the records themselves, which are produced with a subsequent DISPLAY or PRINT command.
- A thesaurus facility should be available.
- The thesaurus contents must be searchable field-level data entry and indexing (i.e., author, publishing date, etc.) must be supported.

On the other hand, CCL has no support for traditional database functionality and RDBMS integration. This is its greatest weakness. There is no mechanism for linking files or concurrently accessing a text and nontext database.

CCL recognizes that the information retrieval environment is diverse and dynamic. Therefore, it does not impose standards and rules on areas that are best served on an implementation-specific basis. Indeed, the introduction to the draft standard states, "The standard does not restrict or prohibit the use of other modes of user system interaction such as menu or natural language interfaces, or the use of a native nonstandard command language along with the standard language." The spirit and purpose of this proposed standard is to provide a basic common language to facilitate the use and implementation of information retrieval among several separate text retrieval systems.

Among the proposed query language standards, CCL is the strongest in the areas of complete query syntax, text retrieval specific functionality, and storage media independence. Its primary focus and concern is the human-to-computer process. Unlike standards such as Z39.50, CCL does not address the computer-to-computer protocol, or how information should be accessed and transferred in a multinode network environment. The primary strength of this proposed standard is as a common lowest-level language between the user and the search engine.

Despite its strengths, this proposed standard has received little visibility in the end-user environment and minimal attention from the

[8]A copy of the Z39.58 standard can be requested from NISO by writing P.O. Box 1056, Bethesda, MD 20817, or calling 301-975-2814.

vendor community. At present, only one product, TRIP from Trip Systems International, is based on this standard.

CD-RDx

CD-RDx is deeply rooted in the overall computer objectives and strategies of the U.S. government. In an effort to increase productivity and streamline costs, various departments of the U.S. government have been looking to EDMS for several years. Indeed, the highly publicized CALS initiative is a prime example of this. In this same spirit, the *Central Intelligence Agency* (CIA) has become a major user of published CD-ROMs. The intelligence community realized early on, however, that the benefits of this publishing media were undercut by the lack of a standard method of text retrieval. For this reason, the *information handling committee* (IHC) of the intelligence community, along with assistance from Helgerson Associates, has designed and is promoting the CD-RDx standard.[9]

The objectives of the standard are: to establish an environment in which the retrieval and the management of data on CD-ROM are both system and software independent; to provide a framework for the production of interoperable CD-ROMs; to promote the storage and exchange of CD-ROM–based information between the various agencies of the US. government; and to deliver a standard that is in compliance with other standards that affect the publishing of CD-ROMs and information exchange in general, such as ISO 10149, ISO 9660 (also known as the High Sierra standard), SGML, OSI, GOSIP, and POSiX.

CD-RDx is not specifically a query language. It is a set of protocols that controls the interaction between the user interface and the database server. These protocols comprised a set of commands, two-dimensional tables, and a procedure for their use. The actual query language is somewhat inconsequential to this standard, and is typically front-ended by any number of user-friendly systems. Thus CD-RDx is a standards contender at two levels of the query architecture, as a common command language (though it is weak in its support of this layer) and as an API protocol.

Key to the standard is the definition of the client and the server, and the role each plays in data indexing, storage, and retrieval. The client may or may not be located on the same platform as the server, which must be physically located with the data on the CD-ROM. The client primarily operates as the user front end. The client handles all commu-

[9]A copy of the CD-RDx standard can be requested from IHS Intelligence Community by writing P.O. Box 90828, Washington, DC 20090.

nications with the user and manages the display of data. The server is responsible for handling the data, including any data compression and expansion. All indexing and retrieval routines are executed on the server. The server is also responsible for managing all communications with the operating system.

When a user enters a query using any interface program available, the client passes the query to the server using the established CD-RDx commands and protocols. The server passes the formatted query to the CD-ROM retrieval program, which processes the query utilizing whatever search algorithm is resident in the database program. If data is found that satisfies the user query, it is passed to the server. The server passes the data to the client, which presents the query response to the user in some type of proprietary format. In this way, whether a user interface is integrated with the CD-ROM database or not, any user interface that complies with the CD-RDx specification can access any data on the CD-ROM. Users are given the option to select the interface that they prefer to use, and are not required to learn the interfaces integrated with each published CD-ROM. This client/server architecture supports backward compatibility. Either the system's servers or clients can be updated and improved without affecting the other.

In addition to supporting content-based query, the CD-RDx standard supports data browsing. Browsing is integrated into the standard since the information being accessed is typically textual in nature. Therefore, the user may want to simply browse the information rather than immediately query it. To provide this capability, CD-RDx supports a structure known as the *table of contents* (TOC). The way in which the TOC is structured and handled is not controlled by the standard. This is left to each CD-ROM publisher. The TOC can be a string of text or a set of tags scattered throughout the full text. In any case, when the user asks to view a table of contents or summary of the database, it must be delivered to the server, which will treat the TOC as just another available index.

CD-RDx is specifically constructed and designed to facilitate the retrieval of information published on CD-ROM. Because published CD-ROMs cannot be updated, there is a lack of support for dynamic databases. It neglects the ability to add, update, or delete text. Indeed, the standard specifically makes mention of this characteristic of CD-ROM, and indicates that the standard was designed with this in mind. The developers of CD-RDx regard the permanency of data on CD-ROM as a benefit, and have taken this into consideration when developing this query and retrieval standard.

CD-RDx views the text query process as one synonymous with the client/server model. This design approach provides interoperability not just between CD-ROMs, but also between any query tool and any indexed database. The scope of CD-RDx, however, is more far-reaching than the scope of the other proposed standards. On one hand, CD-RDx dictates a series of exact protocols and commands to control the data exchange, but it also provides access to any and all indexing schemas and any field structures. CD-RDx takes a black-box approach to text retrieval, in which the primary role of the standard is to act as an invisible translator between the query engine or the client, and the retrieval and/or indexing engine or server.

The blatant omission of updating functionality is a potential hard stop for CD-RDx. If it remains a standard for CD-ROM only, which is all its authors ever intended, updating is not a problem. But many in the text retrieval market are looking to be able to apply this standard to any text database. Text databases are not all static. Additionally, they are not exclusively published on read-only media such as CD-ROM. It would be beneficial to the text retrieval industry in general if potential text retrieval standards supported full database functionality, not just query. Though not specifically defined in SFQL or CCL, the integration of commands that support this functionality is not outside the standard's structure or definition. CD-RDx, however, has been specifically designed with recognition of the fact that updates are not available. This innate direction in design could prove to be a major hurdle to overcome in extending the standard's capabilities.

A similar argument can be raised regarding CD-RDx's lack of a text-specific query language. Though its design facilitates the use of any query language or front end, because it does not specifically contain a query command syntax, there is a lack of direct support for advanced query functionality such as thesaurus usage, synonym files, and concept-based queries. However, there is nothing innate to the proposed standard's design that would preclude the integration of this functionality. Indeed, within its last two updates, CD-RDx has been broadened to include direct support of stopwords, phrase searching, and full proximity.

Despite the handling of objections, and an early acceptance and adoption among several federal government agencies, support for CD-RDx has recently plummeted. It has all but disappeared from the mainstream market, but is still technically a vying standard.

SFQL

Structured full-text query language (SFQL) was originally defined, developed, and promoted by the standards governor of the airline industry, the *Air Transport Association/Aerospace Industries Association* (ATA/AIA). Early in the 1980s, in an effort to reduce paper and increase both accessibility and efficiency of documentation creation, the airline industry adopted CD-ROM as the media for its voluminous aircraft manuals. Each individual CD-ROM manual met the needs of the user and achieved the goals of the project.

Collectively, however, the CD-ROMs from the many individual suppliers and manufacturers for the airline industry created a new problem: lack of interoperability. Each vendor created its respective CD-ROM documentation in a proprietary manner. The system used to read one vendor's manuals could not read those of another. A separate system and set of query tools had to be learned for each manual. It became painfully apparent to the airline industry that a standard was needed that would allow CD-ROM systems to be accessed consistently through a common front end (i.e., query language). This would allow vendors of electronic publishing and retrieval to continue to use proprietary search algorithms, storage strategies, database structures, etc., while still meeting the needs of the end user for a common look and feel to all systems. In 1992, an IEEE committee was formed to handle the promotion of the standard.

SFQL was designed as both a superset and a subset of the RDBMS SQL standard. To understand how SFQL's designers have adapted the SQL language to text databases, a simple analogy can be used. The relational SQL language is built around the concepts of data. It is stored in records that, in turn, are organized into tables or files. The result of a SQL query is a subset of rows in a virtual table. SFQL is based on textual data, organized in one massive table (the database itself) as a series of records (the individual documents). The result of an SFQL query is also a subset of rows (documents, in this case) in a virtual table (a subset of the entire database).

Focusing attention on full-text capability, SFQL does not address or support the relational database capabilities of its parent SQL. It does not encompass SQL's join, update, view, creation, and transaction processing capabilities. SFQL does not explicitly disallow these functions, which lends to its flexibility. Like all text retrieval standards emanating from the CD-ROM market, SFQL focuses on the functionality required in a database of static information. No attention is paid to the functionality required in a more robust text system such as data modification and revision tracking. Thus, these standards will require enhancement if they are to be applied universally to all text databases, not only those on CD-ROM. The effort required to enhance SFQL in this manner should be abetted since it is a subset of SQL that supports and provides functionality to address these needs.

Starting with this subset of SQL functionality, the SFQL design team extended the capabilities of SQL to the realm of full-text retrieval through the incorporation of new commands and options. All new language components were added under the guidelines and syntax rules of SQL to ensure that SFQL would be a SQL-compliant extension of SQL.

SFQL provides more than a standard query language. It also addresses how information is transferred in a network environment. SFQL decouples the search facility from the user application, thus defining a client/server model for text retrieval. Under SFQL, the server software responsible for the actual location and retrieval of text would reside on the CD-ROM in a format proprietary to the individual publisher. These SFQL-compliant systems would understand and enact commands from the SFQL clients (user stations). In this way, SFQL provides interoperability among various proprietary search methodologies on CD-ROMs. This approach to the retrieval environment also positions SFQL as a competing standard in the API layer of the query architecture.

SFQL controls the format of the returned information and the status of each command in a structure known as the SFQL communications area (SFQLCA). SFQLCA consists of a header and a message area. The header is used to hold status information. The message area is used to return data. Under SFQL, the server responds to each command issued by the client with an SFQLCA. In this way the client can interpret the data returned by the server, regardless of how it is stored or handled on the server. The message area is formatted in multiple records of various lengths, which consist of multiple *field subrecords* (FSRs). The FSR is used to identify the source of the data and its type. In this way, SFQL can handle any data type that is defined in an FSR (i.e., fixed-length text, unformatted text, numeric data, and binary objects). Therefore, SFQL controls the text retrieval process at the application, or client end, as well as the application programmer interface end.

Although SFQL was designed to control unstructured information (text), it does impose a degree of structure on the text. The structure is required to support the text in a relational model, which is required by SQL; this is in keeping with the SQL RDBMS metaphor used to illustrate how SFQL's designers have adopted the SQL language to text databases. SFQL is capable of working with fully unstructured text, but it is optimized for text that is tagged. In this way the individual components of a document (i.e., author, title, footnotes, body) are treated as relational-like columns (fields). The query language encourages the use of these field names to limit the scope of a query and control the values returned. For example, a query could be confined to occurrences of the word *Porter* within the AUTHOR field only. A query for all occurrences of the word *cancer* in any part of the document could have its return values limited to just the title fields of those articles.

By recognizing and structuring text in this way, SFQL establishes an infrastructure that will easily play into the RDBMS nature of SQL. The relational concept of views, where individual users of the database could be presented with customized perspectives into the data (i.e., certain user classes see only certain paragraphs of information), could easily be integrated into this environment. Likewise, virtual chapters and documents could be constructed by employing the relational join and intersect functions.

This method of structuring the data places some challenge and limitation on the text database as well. Although a proprietary text structure is not specified, some form of tagging (e.g., via a markup language such as SGML) is necessary. In addition to the overhead associated with tagged text, the user must also be cognizant of the structure or schema of the data in the database. This creates an end-user environment that is less flexible and intuitive than those associated with other proposed standards such as CCL and CD-RDx.

SFQL was generally adopted by the airline industry and has proliferated into many government agencies in the United States and Canada. Work within IEEE and the SQL access group has become somewhat stalled as of late. Arguments regarding the applicability of a relational model to a text environment continue to raise dissension in the market. Although more successful than CCL and CD-RDx, the industry-wide adoption of SFQL is still not imminent. This is despite the support of a handful of text retrieval vendors, the most notable being Fulcrum Technologies.

Z39.50

The origin of Z39.50 is deeply rooted in the library research world. At its inception in the late 1970s, Z39.50 was concerned with linking large bibliographic files maintained at several sites (included in these were the Library of Congress, the Online Computer Library Center, and the Research Libraries Information Network). By the early 1980s, this undertaking evolved into a formal project known as the *Linked Systems Project.* The major players remained the same, but the scope of the standard was altered. It was at this time that Z39.50 was envisioned as an OSI application layer. Through the 1980s, under the direction and efforts of this project team, Z39.50's capabilities were greatly broadened. Its application reached beyond the bibliographic records it set out to manage. The protocol could search a vast array of information types. By the late 1980s, its definition included a structure for the development of clients that could query and retrieve disparate information resources from a series of information servers in a distributed environment. Z39.50 was officially adopted as an ANSI standard in 1988.

This standard continues to be refined and expanded. A committee at the *International Standards Organization* (ISO) is developing a similar standard, based on Z39.50. It is expected that once the ISO standard is completed, Z39.50 will be updated so that it is a fully compatible superset of the ISO protocol.[10]

In large part, the ongoing modification and expansion of the Z39.50 protocol is driven by the community of users that have rallied behind it. For the past five years an official Z39.50 implementor's group has existed. This group monitors and coordinates the efforts of its members and discusses further enhancements and refinements to the standard. There are several major Z39.50 implementation projects actively under way.

Z39.50's approach is far more general and open than the other full-text query language standards (SFQL, CD-RDx, CCL). It is specifically designed to function as an *open systems interconnect* (OSI) application layer protocol. Basically, Z39.50 defaults much of the retrieval functionality to the underlying retrieval engine. There is no inherent database model or retrieval methodology in this standard. Thus, Z39.50 addresses the API protocol layer of the text retrieval query. Technically, Z39.50 does provide a query language syntax. True to the structure of the standard's open architecture, however, the level of detail governing the functionality and syntax of the query language is minimal. In essence, Z39.50 will support any query language that is taught to the Z39.50 back-end processor.

The fundamental model of this standard is that of the client/server. Z39.50's main function is to facilitate the exchange of information, of any type, from any server to any client. There is little definition imposed on the information structure. A database is defined simply as a collection of records. Records are a collection of fields. There is no inherent structure imposed on either of these data elements. Index terms, data types, and data structures are not governed by the standard. While this provides for a more open approach to information retrieval, it does require more work on the part of the implementor. The structures for each of these data properties must be defined by the implementor for each implementation.

[10]A copy of the ANSI standard Z39.50 can be requested from The American National Standard Institute, 1439 Broadway, New York, NY 10018, (212) 642-4900. More information concerning the standard can be obtained by contacting the Z39.50 users group, chaired by Mark Hinnebusch, Florida Center for Library Automation, Suite 320, 2002 NW 13th Street, Gainesville, FL 32609.

Under the Z39.50 protocol, a query is issued on a client. The query is processed on a server, which produces a result set, or a virtual file which identifies the records in the database that satisfy the query. The result set is maintained on the server. The result set does not have to be a copy of the records, but simply a list of references to the records. The standard does not impose any rules on the server regarding record locking, or the maintenance of several active result sets. In other words, there is no control over data update and deletion, and a system that manages the results of only one query at a time is still in compliance with the standard. In practice however, implementors have integrated both record locking and the tracking of several active result sets into the protocol. In these cases the standard does provide multiset handling capability. The client names the result sets (typically defaulting to a sequential numbering scheme) and can delete any or all result sets maintained on the server at any time. In addition the server can automatically delete result sets as needed (e.g., the client exceeds the maximum number of result sets, and therefore the server deletes the oldest of the existing sets).

When a request is successfully completed and a result set is created, the client can request the transfer of any single record contained in the result set. This causes the data record (bibliographic record, text document, image, etc.) to be transferred across the network to the client. Until the client makes such a request, no data is transmitted.

The Z39.50 standard has been implemented and tested in several projects and products. It is the most widely used of the standards discussed in this section. This is due primarily to its acceptance and accelerated utilization on the Internet as a protocol for handling document transfer. Many projects or applications of Z39.50 exist, the most notable being *wide area information access* (WAIS). WAIS is being conducted jointly by Thinking Machines, Dow Jones, KPMG Peat Marwick, and Apple Computer. Since its inception in 1989, WAIS has greatly expanded the original scope of the Z39.50 standard. WAIS is designed more for the retrieval of full-text and multimedia information. The protocol, or means for exchanging the queries and data across the network, is the component that has been least altered in WAIS. The changes in this area are enhancements, not alterations. For example, WAIS specifically supports audio, image, and video data.

In an effort to promote the use of this implementation of the Z39.50 standard, Thinking Machines makes the source code for the WAIS system available over the Internet or by mail. Though the code is free, no support is offered. There are approximately 150 different universities who have developed WAIS servers and clients. There are over 25 WAIS-compliant public databases available over the Internet as well.

The greatest strength of Z39.50 is also its major flaw: its openness. On one hand, there are critics who praise this characteristic of the standard. They claim that only in this way could the standard be so broad-reaching, applicable to a variety of research environments and database models. On the other hand, there are critics who express frustration over the standard's lack of a clear definition for query types, error processing, and file naming conventions. If the author of a Z39.50 system integrates a particular query language into a specific implementation, which in itself is a major undertaking which developers typically attempt to avoid by employing existing standards, the result is a system that promotes interoperability among its own clients and servers, but not with other Z39.50-compliant systems. Clearly, there is a need for more definition and control. Undoubtedly, this will come from the pressures placed by the implementation community, such as the WAIS project and the Internet community in general.

Standard Query—Which Way to Bet?

It must be stressed that while the discussion of the query language standards becomes technical, the end result of implementing such standards is the creation of a transparent end-user interface that will facilitate text application development and user access. However, this will not happen until some order and direction is placed on the chaos that currently shrouds this arena.

If you are wondering how to bet in the meantime, it is safe to say that there is little chance that CCL (Z39.58) or CD-RDx will survive the standards battle. Both proposed standards have failed to rally up enough support to attain a critical mass or a momentum. SFQL was off to a good start and made great inroads when it attracted the attention of user groups beyond its sponsor (the AIA/ATA), and the endorsement of IEEE and the SQL Access Group. But the momentum that was once behind this standard has recently slowed, and one must wonder at what point the user community and vendors alike will begin asking for an ODBC extension (backed by Microsoft) in order to bypass this uphill battle.

The only sure bet is Z39.50. The momentum and critical mass that it has achieved over the last few years because of its use within the Internet community will be hard to derail. But this is still only a piece of the standards puzzle. Z39.50 does not adequately address the common query language layer of this equation. Because it offers functionality at the common query language layer as well as the API layer, it is possible that SFQL will be adopted as the standard at the common level, interfacing to a Z39.50 back end. HyQ is also a contender in this arena. Because SFQL and HyQ have been met with some serious objections, and have been slow recently to rally support, it is possible that a yet undefined approach to this area of functionality will be the ultimate victor. Because the adoption of standards such as these is heavily influenced by the user community, it would behoove you to proactively support either of these standards, if you have a particular bent toward one.

Evaluating Text Retrieval Systems

Given the diverse approaches to text retrieval and the wide array of available functionality, how does one evaluate which approach or methodology is most suitable to a given organization or research environment? There are no simple answers to this question, although one thing is certain: an evaluation of the user environment and document database is just as important to the process of choosing a system as it is to system implementation and design. Before you begin the process of evaluating products, you should become intimately familiar with the environment in which your potential system will be used. Without this prior knowledge, evaluators can waste months of time evaluating products which do not fit their needs. Without taking the time to understand the environment before embarking on a product search, you may ask inappropriate questions, be easily taken in by functionality that will not be necessary and therefore not used, or not recognize functionality that is pertinent.

A common mistake in choosing and designing a document database is to envision a system where the database is designed strictly via the tools of more traditional systems (e.g., normalization in RDBMS), where documents are loaded into the system and the system runs under the friendliest front end available. This approach is sure to result in months of effort implementing a system that may become underutilized or unused due to an inappropriate fit.

The process of evaluating the work environment and database requirements before a text solution is evaluated must include answers to questions such as:

1. Is the textual information an integral part of a larger information database?
2. Do the documents contain images and graphics as well as text?
3. What is the source of the information (i.e., on-line, ASCII documents, word processing documents, hard copy, microfilm, on-line information feeds)?

4. How dynamic is the information? Will changes to the text be necessary?
5. How often and at what rate will new documents be added to the database?
6. How many pages of information are expected to be stored?
7. What is the nature of the language of the documents (i.e., specialized, technical, predictable, prose)?
8. How familiar are the users of the system with the subject matter and the language of the database?
9. How many simultaneous users of the system are there?
10. How is research performed (i.e., reductionary versus abstract)?
11. How is the information used (i.e., lookup, report generation, input to other software packages)?
12. Is this system a new system or an automated approach to an existing system?

Once the responses to these questions are properly understood, a more direct and focused product evaluation process is possible. The evaluator already has an understanding of crucial criteria such as whether an inverted index methodology will work in the particular situation or whether a full-blown thesaurus is necessary. Armed with this knowledge, questions like these should be asked when evaluating a product:

1. Can the system handle fixed-length data as well as text and/or images and can the different data types be handled simultaneously?
2. What text scanners, and OCRs are supported and/or recommended for reading documents into the system?
3. What sort of conversion and preparation of documents is required? Are documents left in their original format (i.e., ASCII or a word processing format) or must they be loaded into the system?
4. How much disk space is necessary to store the index and what, if any, are the extra temporary storage requirements during creation of the index?
5. What are the overall storage requirements?
6. Which compression techniques are utilized to minimize index overhead and/or document storage?
7. What sort of query optimizations are incorporated into the software?

8. What is the length of time for a typical query during peak usage of the system?
9. How is the input of new documents handled?
10. What is the supported breadth of the query language (i.e., boolean logic, proximity searches, prefix and suffix searches, wildcarding, nested logic)?
11. What is the thesaurus capability?
12. How transparent is the actual use of the thesaurus?
13. Is an on-line browser available?
14. Are search terms highlighted?
15. Is there user control of the index (i.e., stopwords added and deleted; user-defined word, sentence, and paragraph delimiters)?
16. Does the query return a set of records that can be narrowed down by a subsequent query?
17. Can the index be viewed and queried?
18. Is a 4GL available that facilitates integration and customization?
19. What standard DBMS features are available? Is there an underlying DBMS or interface to a third-party DBMS?
20. What security options are available?
21. How is integration to other products and systems handled?
22. Are there practical limits to be aware of (e.g., system performance decreases at 1 GB of data)?
23. Can the system interface with on-line data feeds? Are user profiles supported?
24. What does system implementation involve (e.g., stopword list needs to be created, expert rules must be taught to the system)?
25. What type of ongoing maintenance and management is required of a database administrator?

6 Multimedia

Multimedia and EDMS

Multimedia is not new. Ours is a media-based society. Global images and sounds are transmitted to our homes nightly; voice mail has become a fact of life; our children are nurtured by the glow from video games and transported into virtual worlds. Why then all the hype and excitement about multimedia? Is this really the next tidal wave of computing, or just another ebb in the vast ocean of change sweeping this industry? If you ask developers and users this question answers run the gamut from multimedia as the pinnacle of computer technology, to multimedia as the nemesis threatening to bring computers and networks to their knees. The answer, as is often the case, lies somewhere between the two extremes. For information professionals, the real question is not so much a philosophical one as it is a genuine concern for the challenges of integrating this new technology with existing investments in IS. How do we build multimedia systems that enhance the organization's systems investments rather than displace them?

Multimedia technology is quickly evolving and will just as quickly capture the interest of end users. These are end users with just enough desktop power to turn the effects of this ripple into a tsunami that threatens to wash away the foundation of the information systems that we have built over the past 10 years. What is needed is an understanding of the context of multimedia's role in the evolution of information systems and what we as IS professionals can do to develop bridges that join existing systems with the potential benefits of multimedia.

Multimedia's Role in the Single Point of Access

The first step is an appreciation of the role multimedia plays in the larger trend of single-point-of-research technologies. In the single-point-of-research environment the computer becomes the interactive vehicle for the delivery of all information types. The user has access to any data type through a single application interface that is coupled with a comprehensive database. Viewed in this way, many contemporary multimedia solutions fall short.

Engineering true multimedia applications involves more than simply capturing and displaying vastly different data types. Information arriving in various forms from multiple sources must be carefully orchestrated, allowances made for interactivity, and a solid structure built for the immediate retrieval of all data types based on their content. Without these essential functions, the result is not multimedia, but polymedia—the computer user's equivalent of the couch potato.

Single-point-of-research technology, like multimedia, creates a singular conduit for the transfer of all digital information. Rather than rely on several information sources which exist in separate physical repositories, all the information necessary to support an activity or a decision is available through the single point of research.

What is often ignored in most multimedia applications is the importance of the database foundation upon which single-point-of-research multimedia is built. It is the database that ultimately provides the benefits of the single-point-of-research model. And this, in turn, provides the foundation for integrating multimedia with existing information systems. Although there are many applications of multimedia that do not rely on a database structure, combining multimedia with single point of research results in an encompassing information environment that provides benefits far beyond those of multimedia alone. So how do you build a multimedia solution that supports a single-point-of-research environment? There are many aspects to this, but the one I will focus on here is the multimedia database.

The Multimedia Database

It's a common assumption that powerful hardware is at the heart of multimedia. Hardware power is indeed necessary, but after we lift away the layers of interface and applications, at the very core of every effective multimedia environment is the database. It's funny how valuable this relic of third and fourth generation technology becomes. The database is the one constant underpinning that binds together almost all information technologies. With multimedia, storage mechanics, entity relationships, and retrieval methods play large roles. Not all databases handle all of these multimedia requirements well. In fact, few handle any of them.

In its simplest form, storage of a multimedia data type implies the ability to deal with large data objects, such as voice or image. This means that the database must either store a pointer to the data object that exists as a separate file, or the database must actually store the data in a database field. The latter approach has been commonly known as a *binary large object,* or BLOB. The appeal of BLOBs is that all of the information resides under the control of the DBMS. By definition, security and manipulation of the data occur through the tools familiar to the application developer. Thus, multimedia databases provide the single-point-of-access model the ability to track and deliver another level of undistilled information. The ability to digitize video and sound makes it possible to store true-form information of this nature, in the same way that imaging provides true-form access to pictures. Therein lies one of the major benefits to the inclusion of multimedia in an EDMS, although the problem of distilled access still persists in the case of multimedia data.

BLOB fields are primarily repositories of information, and so are not necessarily capable of presenting or acting on the information. For example, a BLOB that stores a large body of text may have no mechanism for retrieving that text based on words or character strings. It is capable only of taking the request for a procedure and passing the text to an application. At a minimum, the BLOB should provide for intelligent storage of the information it contains: e.g., a video data screen would be passed to an application in such a way as to meet the demands for a particular frame rate or a specific sequence of frames from the video. If the BLOB is unable to do this, serious inefficiencies will occur in the process of retrieving the information, since the application must always work with the entire BLOB. This is a minimal threshold of intelligence that occurs at a low level, transparent to the application. At a higher level, the DBMS must now support the direct requirements of the end user, the most important one being the retrieval of multimedia objects.

Retrieval of Multimedia Objects

Retrieval of multimedia objects is perhaps the single most difficult and unimplemented aspect of the technology. Although a great deal of research and some promising technology exists in this area, almost none of it has reached the market. The only notable exception is text retrieval (refer to Chap. 5 for more information). In a multimedia application that contains text as part of compound documents with other image, video, or voice types, links can be constructed among different objects and the text is then used as the driver for retrieval. This is a powerful approach but it does not recognize the requirement to retrieve documents based on the content of an image, or a video clip, or a sound bite. It is entirely possible that a user may wish to retrieve an image based on another image, or a full-length video based on several frames of that video, or the same video based on a sound bite.

The problem with text-based searches in multimedia databases is that they are highly ambiguous, even if you have taken great pains to carefully index the data and prepare the search request. If you rely on text to retrieve multimedia data types you are relying on second generation information, distilled into descriptive text from the original. Consider the scenario on looking for the inaugural speech based on the phrase "Ask not what your country can do for you." That may seem to be a trivial process. Yet if you had a multimedia library of all modern-day political speeches, which had been transcribed as text, and issued that request, the result would be an enormous number of speeches, since many have used the phrase since it was made famous by President John F. Kennedy (JFK). On the other hand, if you had a sound bite of the speech recorded out of the original context, it could easily be submitted as a retrieval clue for a content-based search.

The pattern recognition technology discussed as a text retrieval search methodology in this book can be extended to include the content-based retrieval of any data type stored in digital format. Thus the retrieval of JFK's speech based on a sound bite is possible. But, while this holds great promise, in most cases, practicality dictates retrieval based on the textual element of the compound document. Within three years time, the implementation of this technology should become much more commonplace. It is already being used for battlefield recognition systems that identify enemy troops and armament. With such a system, the user could bring up image, video, or voice extracts of a data type and submit these to the multimedia retrieval engine. The result would be the ability to retrieve multimedia database objects on their contents. We could then apply many of the traditional DBMS retrieval metaphors such as query by example (QBE), boolean query (AND, OR, NOT, XOR), and content-based query.

Entity Relationships. Finally, there is the issue of multimedia DBMS entity relationships. In traditional DBMS technology, data entities are related by virtue of shared data items as with a relational database, by predefined links, or with an object-oriented system through rules-based links. In a multimedia environment, all three of these constructs can be used. The difference, however, falls in two distinct features of multimedia databases: synchronization and binary object links. These aspects of the multimedia system are difficult to overcome. The first, synchronization, involves the timing involved to present more than one data type at the same time.

Earlier we said that the difference between polymedia and multimedia was this important feature of orchestrating multiple data types with an interactive environment. The difficulty sets in when we attempt to create a database that will preserve this type of synchronization. Traditional databases simply do not have mechanisms in place to provide for this. A multimedia database that does not ensure the integrity of concurrent data types such as video and voice is unlikely to establish a high credibility or utility with users. Stop and think about a poorly dubbed foreign movie where the movement of the actor's lips and the words spoken seem to have little to do with each other. The fact is we have become intolerant of visual media that do not meet very high quality standards.

The Challenges of Synchronization. Synchronization also refers to the importance of playing the same multimedia presentation on several platforms of varying performance. This also does not present a problem for traditional alphanumeric information. The presentation of structured data is not time dependent. In other words, an element of alphanumeric information is either displayed or not. It is never an in-process event. Although you may wait longer for a report to print at a slower machine, the integrity of the report does not change by virtue of this inconvenience. Multimedia information is very much time dependent. Audio or video at three-quarters speed, or, worse, audio at three-quarters speed and the accompanying video at three-eighths speed, impacts directly on the integrity and value of the information.

There is no perfect answer to this, but there are ways to mitigate the problem by adjusting the amount of information presented. A multimedia database may, for example, provide every other frame of video if the machine is half as slow as the machine on which the multimedia presentation was developed. If it has a sound card on board, sound could be provided at full throttle while the video is compensated by this same method. Whatever the specifics, the database must play an integral part in this process to preserve the integrity of the multimedia data. The role of the database is expanded in these scenarios to that of an object manager that also contains the information necessary to correct, manipulate, and present the various multimedia objects.

Binary object links are also part of establishing entity relationships. A binary object link is the equivalent of joining tables in a relational DBMS. In a multimedia DBMS, however, the table items are much larger and contain links to other objects based on the principle of synchronization, which creates instances of links not just from one binary object to another but also within a binary object. To visualize this, let's go back to the video of JFK's inaugural address. The multimedia version might have an option for younger viewers to see some of JFK's accomplishments while in office. This may be a quick segue to a video of the Peace Corps at the point in the speech when JFK says, "Ask what you can do...." That requires a link from one binary object, the inaugural address, to another binary object, the Peace Corps video. These interobject connections need to be preserved somewhere in the database for the purposes of integrity, maintenance, and flexibility. If, instead, they are hard wired, as they might be with a simple hypermedia system, the effort to author, modify, and update these links becomes an enormous burden for the multimedia system and its developers.

Why Multimedia?

Aside from the benefits of eliminating data distillation and enhancing the single-point-of-research model, there is another underlying and related reason why organizations are embarking on multimedia applications. Multimedia is a natural conduit of information for a generation raised on television. We process information at a high bandwidth and can assimilate multiple information channels simultaneously. This is ultimately what multimedia is all about. When our senses and mind are interactively "entertained," we tend to learn faster and better. A study done at Stanford University found that we typically retain 20 percent of what we hear, 30 percent of what we see, and 60 percent of what we interact with. Multimedia, by providing an interactive environment that engages sight and sound, raises the computer's ability to communicate significantly. This is the bottom line reason for the investment in multimedia despite current entry-level costs, processing and storage overhead, lack of precedent, and the level of in-house expertise required.

Examples abound, from simple CD-ROM-based publications to mainframe-based applications. Bristol Myers Squibb, for example, distributed a survey on an interactive multimedia CD-ROM. They found that they received 30 percent more responses than was typical for similar paper-based surveys. More importantly, respondents provided better input, spending approximately 25 percent more time in filling out the survey. When asked why, respondents indicated that the survey vehicle was *fun*.

The United States Postal Service invested millions in a multimedia system that was integrated with their barcode reading machines. The multimedia system trains postal employees on using and servicing the equipment. By providing the worker with *just-in-time training*, they were able to recoup costs by eliminating the need for costly computer repairs by third parties. (If the barcode reader malfunctioned, it would invoke the multimedia machine which would begin a problem-solving and/or training session with the operator. This on-line book included full-motion video, sound bites, text, and images.) Similarly, IBM reported a cut in internal training costs by $300 million annually by migrating to a multimedia on-line training system.

Florsheim Shoes increased sales by 20 percent at 500 locations across the United States by locating a multimedia kiosk at each of these locations. Shoppers could interact with the system, receiving video- and sound-based information in an interactive session. If the shopper wanted to ultimately purchase product, the system accepted credit card payment and would process an order to ship the shoes to the customer's home address overnight.

The city of Atlanta, Georgia, spent $1.8 million and utilized literally every computer resource at Georgia Tech to produce a multimedia presentation to the Olympic Committee. It resulted in Atlanta's being chosen as the host city for the 1996 Summer Games. It is estimated that the $1.8 million investment will render $3 billion to the local economy. Investments in multimedia can be high, but the payback can also be commensurate.

Because of its overt connection to education, the multimedia market is predominantly that of education-based application—approximately 61 percent of all installations. Information kiosks, such as the ones developed by Florsheim, account for another 38 percent of the market. This market segment, however, is the fastest growing. Business communications, such as the presentation staged by the city of Atlanta, and general publishing applications each account for approximately 1 percent of the market.

Although the educational market is best known, multimedia developers estimate that the business market is potentially two to three times that of education and the consumer market is potentially 20 to 30 times the size of the business market. Total market size is projected to $10 billion by 1996 with the consumer player segment representing the single largest share.

The Multimedia Challenge

You are faced with multiple challenges in migrating to a multimedia-enhanced EDMS. Challenges exist on two plateaus: creation of applications and delivery of applications. From the start you must recognize that the role of a multimedia programmer is very different from that of a traditional database programmer. With the advent of new data types comes demand for new skills. Multimedia applications require personnel to handle overall project management (a producer), interface design, content expertise, instructional expertise, and media expertise (i.e., composers of music, sound specialists, and movie editors). The good news is that advancements in multimedia authoring systems have eliminated the need for many of these by automating and facilitating functions such as sound production, editing, etc. But, do not be fooled. If you embark on a state-of-the-art, full-fledged multimedia production such as that created by the city of Atlanta, you will undoubtedly find yourself hiring a staff of musicians, producers, editors, etc. Today's tools make it possible for a single programmer to produce relatively simple and short multimedia applications.

Aside from the personnel issue there is the matter of authoring platforms. For many organizations this represents a major investment, as much of the equipment necessary to empower the multimedia author is not in-house. An entry-level authoring station can cost approximately $10,000. (See Fig. 6-1.)

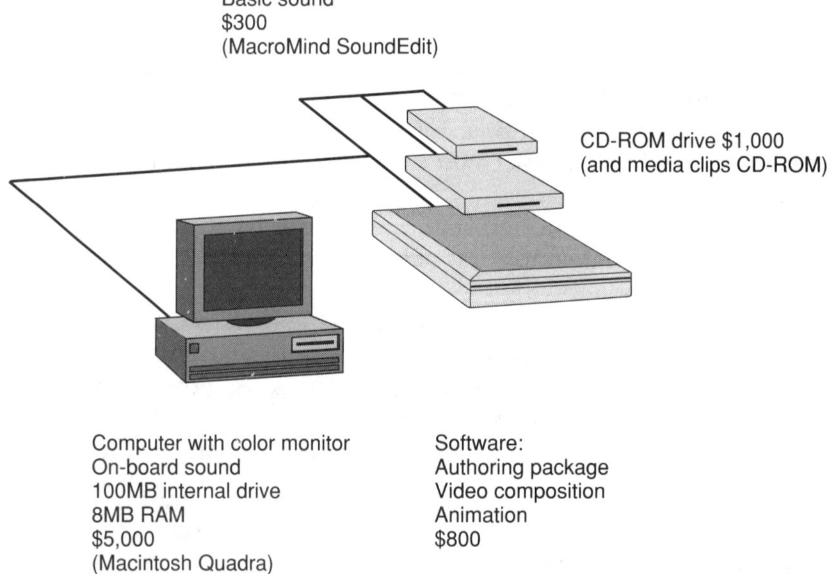

Figure 6-1. A basic-level multimedia development platform. A basic authoring system consisting of low-level authoring software, a workstation configured with a color monitor and sufficient processing capabilities, a color scanner, and a CD-ROM drive will cost approximately $8700.

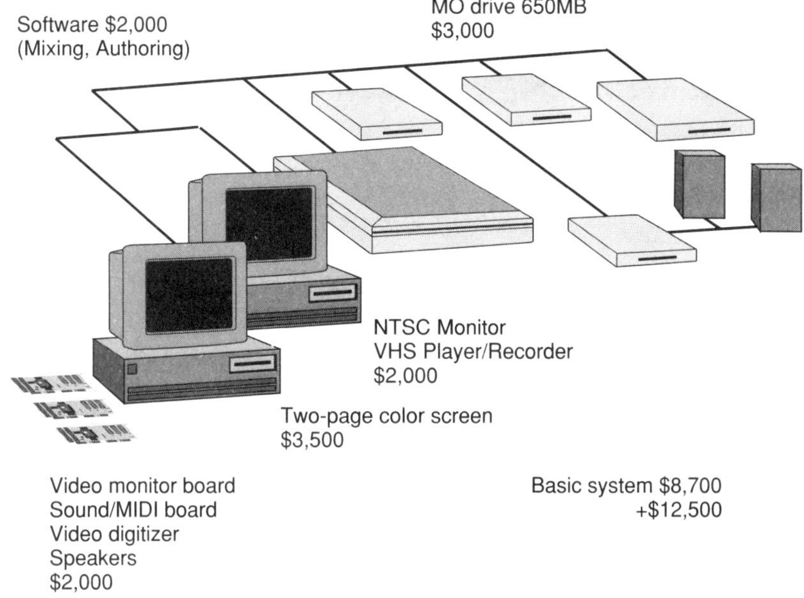

Figure 6-2. Midlevel multimedia development platform. The midlevel multimedia authoring workstation provides enhanced video and sound capabilities. Storage capacity is increased to accommodate their usage. Additionally the author is given better resolution and larger viewing screens to facilitate editing.

A basic system, however, inhibits editing by providing a single viewing station and not providing sound mixing and authoring capabilities. By providing these capabilities along with additional storage capacity and enhanced video, the authoring system's price tag leaps to more than double that of a basic system. (See Fig. 6-2.)

For state-of-the-art productions, however, additional memory, input facilities, and processing power are required. (See Fig. 6-3.)

You must keep in mind that these scenarios are for a single-user workstation. Overall system costs will obviously be significantly increased with each author empowered. Among other reasons, the reality of system costs have kept many organizations from launching an aggressive push to multimedia.

The inclusion of additional storage in each of the scenarios should not be taken lightly. A serious consideration in incorporating multimedia into the EDMS is the significant increase in storage that it entails. This is true even in cases where multimedia will be minimally used, such as voice-annotating documents. The reason is the innate file sizes associated with voice and video data types. As was the case with

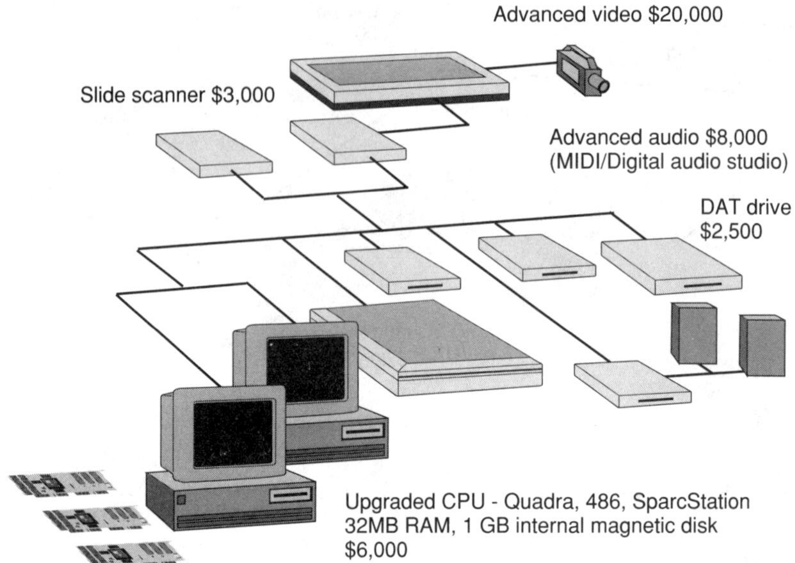

Figure 6-3. High-level multimedia development platform. A high-level multimedia authoring platform includes color slide scanning capability, an upgraded workstation with a greater processing capacity, support for direct video input and/or digitizing, and a state-of-the-art sound system. This system configuration costs approximately $60,000.

images, multimedia files demand powerful file compression. Yet, even with the availability of such compression algorithms, incurred overhead can be as high as 1 MB for a second of full-motion video, without sound. Table 6-1 provides an overview of the storage requirements for audio and video. Compression becomes vitally important when you consider that these files must be transported across a network and/or played back from devices such as CD-ROM, which has a data transfer rate of 1.2 MB/second.

Table 6-1. Sample Storage Requirements for Audio and Video

	Uncompressed	Compressed
Audio (1 second)		
CD audio (mono)	86 KB	14 KB
Video (1 second)	55,000 KB	1000 KB

Related Standards. The compression algorithms used in multimedia are among a growing number of multimedia standards. Although not adopted, there are two proposed standards for the definition of a multimedia platform itself. These definitions address the end-user platform, as opposed to the authoring workstations. The multimedia PC (MPC) standard is the product of the Multimedia Marketing Council founded by Tandy and Microsoft. The MPC standard defined a minimal configuration as a DOS-based 386 PC @ 20 MHz, with 4 MB RAM, an SVGA monitor, and a CD-ROM drive.

But the MPC standard is not supported by some industry leaders such as IBM, Apple, Commodore, and Sony. These vendors, along with hundreds of others, are members of the *Interactive Multimedia Association* (IMA), founded in 1988, and dedicated to accelerating the adoption of multimedia through the sponsorship of standards. The IMA defines the multimedia platform as any hardware, operating system, and API that provides multimedia capability. IMA member IBM defines the minimal end-user workstation, known as *ultimedia,* as a 386-based processor, with 16-bit audio, multimedia extensions, and a CD-ROM XA drive.

The *Motion Picture Expert's Group* (MPEG) is the only proposed standard for video file compression. MPEG reduces the size of video files by coding interframe redundancy (eliminating those graphical elements that do not change from frame to frame). This feature makes MPEG up to 250 percent more effective than JPEG. (For more information on JPEG see the imaging standards section of Chap. 4.) The MPEG standard, however, is not finalized. Until such time that it is, many developers are turning to proprietary compression methods such as Apple's QuickTime.

The *Music in Information Processing Standards* (MIPS) Committee at ANSI is developing a standard known as X3.749-D, or HyTime. It expands on the SGML language for linking, time scheduling, and synchronization in documents. This standard is intended to foster the growth of multimedia applications by recognizing the special characteristics of data like music and video and by providing structure to facilitate their integration.

The advent of multimedia has also fostered a number of CD-ROM-based standards including CD-I and CD-ROM XA. *Compact disc interactive* (CD-I) is a proprietary technology developed by Philips. CD-I was developed as a general consumer market platform. CD-I is a self-contained multimedia platform. Another standard, *eXtended Architecture* (CD-ROM XA) is a CD-ROM-compliant format that incorporates support for the storage of video and sound on the disc. Unlike CD-I, CD-ROM XA discs can be played on a CD-ROM drive equipped with an upgrade card.

A Foundation for Change

All in all, it is clear that the database will not only provide a foundation for the complexity of the new multimedia data types, it will also enable the creation of a single-point-of-research environment which will support the flexible manipulation and retrieval of multimedia information. In time this will also provide the security for an effective migration of the benefits of multimedia. For the time being, the reality is just as clear. In spite of all the hoopla that abounds, the demand for multimedia applications has until now been relatively weak. Interest alone is a falsification of actual budget allocated to multimedia solutions. Databases in use today reflect the vast predominance in the availability of structured data in alphanumeric form over multimedia data.

In the short term, you should plan for a progression to pointer-based methods of database access to multimedia data types. These are already available through database vendors that support BLOB fields. Few vendors, on the other hand, have seen the need to migrate beyond these basic functionalities—relegating today's multimedia applications to the periphery of mainstream IS. Interest in integrating a richer variety of data types and sources with these core applications will continue to mount rapidly. As it does, users will voice their requirements for multimedia functionality in basic applications such as E-mail and document management. These will fuel the development of single-point-of-research environments that augment existing database applications and data structures.

If the past few years have taught us anything, it is that new technologies such as multimedia do not wash away existing systems; instead they rely on them as a familiar and stable foundation. Multimedia is one more step toward expanding the database to a fully digital single-point-of-access environment.

7

Designing an EDMS

A New Perspective

Implementing an effective EDMS is as much a matter of adopting a new perspective of your organization as it is understanding the technology. With an understanding of both, you are ready to put in place a revolutionary new tool for office automation. In this text we will review a set of methods for the analysis of an organization's document management and information retrieval requirements. The ultimate benefit of these methods is a new perspective of the organization. It will be a subtle shift in some cases, but at times the shift will be very dramatic, such as the Stair Step methodology—a tactical approach that seems to be entirely contrary to popular belief in reengineering. Adopting this new perspective begins with a subtle twist on the basics of systems analysis, but it will quickly evolve into a new set of analytical tools developed specifically for EDMS applications.

To begin, however, we have to set some objectives. The basic objective of any new information system is the automation of tasks and processes in order to increase an organization's productivity in working with its information resources. In the case of EDMS, this fundamental objective remains unchanged. However, the advent of complex work environments, increased dependency on networked workgroups, and the physical nature of the electronic document require that we revisit existing assumptions about structured information systems and current methods of analysis and design. The considerations, methodologies, and analytical techniques that are presented in this text have been developed to address the specific nuances of implementing an EDMS. The focus of our effort will be on defining the role of workflow in that design effort. This will act as the foundation for the EDMS.

Your workflow forms the overall environment for the new information system. The underlying components of a workflow system are already in place in most organizations. These are most typically office automation tools such as word processing, electronic publishing, spreadsheets, and E-mail. Workflow brings together these existing technologies in an integrated and easily managed environment.

It is helpful to think of workflow's role as that of an extended and distributed database. The database organizes, integrates, and controls data items that could well exist outside the database, but are better managed within its structure. Although traditional database repositories are powerful tools for the organization of structured information, most cannot fulfill the role of a workflow system or an EDMS in general for two basic reasons: they are a poor fit for unstructured document-based information, and they do not provide an environment for the integration of office automation applications.

Traditional databases lack the tools to track and manage the flow of documents. Although these tools can be created with advanced languages and object-oriented programming, EDMS solutions, especially those that focus on workflow, provide tools for scripting a series of tasks that are specifically designed for document environments. This allows for quick application development and process redesign. Both of these will be discussed later in the text.

The first step in implementing EDMS is developing an understanding of the obstacles that implementors face. These revolve around the three primary areas of infrastructure integration, process redesign, and organizational or human factors. Each of these areas accounts for a significant component of EDMS analysis. Infrastructure integration will be discussed in the section on the System Schematic, a graphical tool for depicting the existing and planned information systems architecture that will be used to support the EDMS environment. Process redesign will be addressed in several sections: the System Schematic, reengineering, and Time-Based Analysis. Finally, we will look at organizational issues and human factors in both the Stair Step model and Time-Based Analysis.

In my experience consulting with organizations embarking on an EDMS implementation, I have generally found that organizations can easily grasp the concept of imaging and appreciate, at least at a high level, the capabilities and positioning of a text retrieval system into their EDMS strategy. The same is not true for workflow for a variety of reasons. Therefore, it is beneficial that we take some time now to develop a better understanding of the role that workflow plays in the overall process of information systems reengineering and the specific problems it is intended to solve, before attacking the problem.

Since workflow is often associated and, in many cases, incorrectly confused with reengineering, it is important to first differentiate the two. Although both share some common ground there are distinct differences. Among these is the fact that workflow can, in many cases, be implemented with significant benefit without redefining and disrupting the entire enterprise. Reengineering is often a much more radical approach to deassembling and then reconstructing the entire enterprise. The principle behind this comprehensive restructuring is that the full benefit of reengineering can be realized only if the fundamental mechanics and inefficiencies of the enterprise are removed or replaced with new building blocks. Anything less is usually referred to as "paving the cow paths" by many diehard students of reengineering. But this is unnecessarily dramatic and in many cases simply impractical. Reengineering on an enterprise scale assumes that the organization is stable enough to ride through potentially difficult transitions in relatively short periods of time. Although the case can be made that some organizations are desperately in need of this type of overhaul, it does not come without a commensurate degree of risk.

Workflow does not require drastic change. It is most often applied to smaller workgroups and then scaled up to the rest of the organization over time—not unlike the evolution of many of its component office automation technologies. The approach is certainly less disruptive, but in the case of workflow, it is based on a deeper principle than just convenience of implementation. Since workflow does not alter the basic structure of the information systems used within an enterprise, the office automation, data, documents, and existing IS applications remain intact while they are integrated into the workflow environment.

The Delphi workflow analysis and design methodologies recognize this fundamental difference and benefit of workflow. The methods described in the following sections consist of three separate phases: defining the existing business process and technology infrastructure, identifying the areas of weakness and inefficiency in the definition, and collapsing the business cycle and developing new process models. Each of these steps is a critical aspect of effectively applying workflow and its brethren EDMS technologies.

The first phase, defining the existing business process and technology infrastructure, is accomplished through a technique referred to as the System Schematic. The System Schematic provides a comprehensive definition of the current components that make up the tasks and information systems used in the organization prior to the application of EDMS and workflow. It uses a graphic representation of the present system's technology infrastructure and the flow of information within the present structure. This provides a clear framework for the discussion and examination of problems and alternative solutions.

The second phase identifies the areas of weakness and inefficiency. It uses the results of the System Schematic along with a technique for establishing the timing of each business cycle to target areas of potential productivity improvements. This methodology is called *Time-Based Analysis*. It focuses on the element of timing, which is absolutely essential to proper workflow analysis and the most often ignored aspect of many approaches that use traditional data flow modeling techniques. By using Time-Based Analysis, it is possible to achieve quantum gains in productivity without massive reengineering.

The third phase, development of new process models, uses an enterprise modeling technique called Stair Step. The Stair Step model provides a context for identifying the most appropriate sequence of precisely defined workgroup applications, the first of which will be the pilot application. It stresses an incremental approach to developing enterprise workflow systems. Stair Step takes into consideration the complexity of large workflow applications and the dynamic nature of most enterprises. It recognizes that expectations must be regulated and results quickly demonstrated for the technology to gain favor among both users and management. Ultimately, Stair Step provides a foundation of experience on which to build enterprise applications; today, this is the most difficult ingredient of all to find for any EDMS application and workflow in particular.

The System Schematic

As with most automated systems, the first step in designing an EDMS solution begins with an understanding of the work environment, the work processes, and users' needs and requirements.

It is advisable at this point to have a basic understanding of EDMS systems, which includes a repertoire of the available technologies and products, methodologies, applications, benefits, and limitations. We will assume that the reader has such an understanding from this and other sources.

The System Schematic is the foundation of a well-designed EDMS application. It will be used throughout the entire analysis and design process to assist in understanding how information flows through the organization, how it is processed and accessed, and how the proposed EDMS and workflow environment will be supported by existing or planned hardware, software, and communications infrastructure. The System Schematic is constructed from information gathered by interviewing key individuals participating in the evaluation of the EDMS and users of existing information systems to be integrated into the EDMS environment. These interviews are not intended to determine the specific needs and requirements of users for the EDMS system. This will be done in phase 2 of the System Schematic. The System

Schematic is focused on developing a better understanding of the existing technology infrastructure.

The primary objective of the System Schematic is the development of a common understanding of the organization's existing technology infrastructure among users, evaluators, and implementors. We will use the term infrastructure to denote the full range of all hardware, communications, and software that make up an organization's information systems. The secondary objective is the development of a framework that identifies the major areas of concern and potential difficulty in the EDMS implementation. These may be found in the lack of sufficient communications between distributed nodes, or the inability to support the variety of end-user platforms in place. The System Schematic is, at first, an orientation tool that can set general expectations about the scope of the project and the degree to which existing information systems will support desired results. In its final rendering the System Schematic acts as the foundation for establishing *Request for Proposal* (RFP) requirements and the development of a project plan.

One of the most interesting aspects of the System Schematic is how few organizations already have one in place. For the last four years I have been polling audiences at Delphi's seminars to find out how many of them have a complete and up-to-date schematic of their organization's hardware, software, and IS infrastructure readily available so that they could produce it within one hour if I were to walk into their office and request it. Amazing as it may seem, less than 2 percent of the attendees have said yes. The remainder simply do not have such a road map of their organization. Since an EDMS implementation can have an enormous impact on existing applications and will certainly be affected by the availability of existing systems and platforms, it is an ideal opportunity to develop this type of diagram for the evaluation of the EDMS and the benefit of the entire IS planning process.

The initial System Schematic need not be elaborate. It should serve as a framework for future discussions about infrastructure. There is no need at this early stage to provide extensive detail about the nature of the information and the individual concerns regarding how the information will be retrieved; this will be discussed later in the chapter. Developing the System Schematic is a three-phased process that begins with the definition of a project scope and an assessment of the existing infrastructure.

Phase 1: Defining the Project Scope. The project scope can be as broad as solving an entire organization's document management needs or as narrow as the development of an application to automate a single workgroup. The best way to define the organization for which the System Schematic is developed is to ask what portion of the organization's IS architecture falls under the highest-level corporate sponsor for this EDMS implementation. Identifying a sponsor is critical to the success of an EDMS. The sponsor should be someone that is not only committed to the EDMS application but also has the budget or the ability to justify EDMS to a budget authority. This provides a practical sense of the scope that can be managed for the purposes of correctly defining and controlling the present and future infrastructure.

Without this level of sponsorship it will be next to impossible to gather sufficient momentum or build a substantive case for EDMS. The reason for this is the intangible nature of an EDMS' benefits in many cases. Often the EDMS results in faster access to data, better decisions being made, or other forms of increased productivity that are not easily justified prior to implementation. You must therefore gain sponsorship and implement within an organizational scope that will provide adequate time to demonstrate the value of the EDMS. This piece of the organization will be the principal target for the pilot EDMS implementation. It will not, however, become an exclusive application that ignores the remainder of the enterprise and users outside of the initial project scope. We will attempt to balance the requirements of the pilot application with the overall requirements of the enterprise through the System Schematic and the Stair Step methodology.

The sponsor must also be able to provide clear and succinct *critical success factors* (CSFs). Because the benefits of an EDMS are diverse and the technology can be applied to many different ends, it is necessary that you be given a metric by which you can judge the applicability of technology and its success. Why is the organization seeking change? Certainly there is a reason. Short of an occasional masochistic personality, no one enjoys change. Therefore there is usually a good reason why it is being sought. The sponsor must be able to answer the question "What are you trying to achieve?" and the answer must be measurable. For example, *increased productivity* and *becoming the leaders in customer support* are not CSFs. Virtually every organization would like to increase productivity and become the leader in customer satisfaction. These are vacuous statements. You must insist that the sponsor define these noble goals. Though you may not realize it initially, they can be defined differently. For example, I worked with a software development company who defined being the leader in customer satisfaction as the ability to close 90 percent of all customer calls on the initial call. A West Coast bank that was also a client of mine defined being the leader in customer satisfaction as a 40 percent increase in the amount of follow-up business they do with their existing base. Now, we could argue the logic of their definitions all day if we care to—but I don't; there is no reason to. It is the sponsor's project and goal that you are trying to legitimize through technology application. The definition can be anything the sponsor wants it to be, as long as it is a tangible, measurable CSF.

In all cases, however, the process for creating the schematic is the same. The first step is to define a system architecture for the hardware, networks, and applications that compose the available environment for the EDMS system. This will serve as the general framework for the System Schematic. At first glance, you may assume that this part of the process is an objective analysis, and therefore requires only minimal interviewing—several targeted IS/MIS and key users will typically provide a satisfactory level of detail. This may be the case in theory, but in practice it is not. Because you are constructing a document-intensive business process system, it is a foregone conclusion that you will potentially affect each and every person in the organization. It is therefore important that you gain an appreciation of the entire organization's infrastructure, not only that portion under the direct control of IS. For example, I can recall one client whose IS department stated emphatically that they were an IBM shop—a series of mainframes and PCs linked together. In speaking with various user departments, however, the presence of several Macintosh LANS was uncovered. These were systems that were brought in through departmental budget. Although not technically part of the organization's "official" infrastructure, they were nonetheless work environments that needed to be supported. In fact, many of the documents and business processes that were being scrutinized were generated on these Macintosh systems. But gathering the information regarding infrastructure is only part of the task. Before proceeding to phase 2 one additional aspect of the basic System Schematic must be considered: routing.

Routing identifies the types of information sources, the way in which this information is used, and the path it takes through its life cycle. At this stage of evaluation, routing definition is concerned only with the existing documents. It does not consider any potential changes in the way the information is handled. In most cases this will require expanding the System Schematic to include non-computer-based tasks and manual or semimanual document management activities. The completed routing definition and the initial System Schematic are now used as the primary vehicles for the identification of specific user and application requirements.

At this stage, the Systems Schematic is far from complete and is in fact almost always going to be a misleading representation of the actual and final version. You should resist the temptation to finalize the

System Schematic at this time. If you attempt to finalize it you will make numerous assumptions and subjective analysis that will be based on incomplete information. When we discuss phase 3 we will return to the System Schematic in order to correct the many flaws and incorrect assumptions that have been made so far.

Phase 2: Collecting User Requirements. The next step in developing the System Schematic is the user interview process. Interviewing users directly will certainly offer the best insight into their work environment. Without adequate preparation, however, the process will be frustrating for both the users and the system designer. Without full disclosure from the users, you run the risk of having to repeat the interview process several times before developing a complete System Schematic. It is also likely that users will, at best, tire of the repetition. At worst, you will lose their confidence in your ability to appreciate their needs. This last point is a risk of any systems analysis. It is particularly damaging to workflow analysis, however, due to the level of user buy-in required to successfully implement a system.

Users must not only be a significant source of information for the analysis, but they must also believe that they are a strong contributing influence. Without this sense of teamwork, you run the risk of creating a workflow application that is often circumvented. It is not unusual to encounter organizations using poorly designed workflow applications, where documents are routinely printed, used, and filed in paper form, resulting in minimal benefit from the workflow application.

The goal of the designer-interviewer is to identify weak areas or potential points for collapsing current business processes. It is not uncommon to ask the users-interviewees where they perceive problems and where they would like to see change. The answers provided, however, may identify only symptoms of these problems. Your job as the interviewer is to probe beneath the surface by watching for common threads to emerge in the System Schematic. These may speak to a particular aspect of the existing system that is inhibiting access to information, such as the lack of adequate network bandwidth for transferring documents or the inability to track document status when a document is transmitted electronically. Users in many cases will not directly identify this problem, because of their acceptance of the existing system's parameters. Interestingly, they will usually acknowledge it when presented with such a finding, but confess to not believing it was something that could have been changed.

The interview process at this point is still dealing at a high level, with personnel who are not necessarily involved with the day-to-day intricacies of the day-to-day business. Many users feel the need to respond to questions about the development of new computer-based systems with technical observations about the current system's deficiencies—not unlike the tendency most people have to use a limited medical vocabulary when talking to their physician. This limits the value of the user's observations by confining their perceptions to what they believe is possible. In the context of traditional systems analysis this is actually an advantage since users and developers can establish a common rapport and expectation set. In the context of a workflow and EDMS analysis, however, user presumptions based on existing knowledge of what systems can or cannot do is restrictive and counterproductive because it ignores the benefit of changing work models in new ways that have no counterpart in existing systems.

One technique that is helpful with users who are reluctant to offer an opinion based on their lack of technical know-how is the *blue sky* method. Users are asked to set aside their preconceptions about information systems and describe an ideal scenario. For example, one user, when asked to describe how to improve on a customer support system replied:

> It would be nice if I could just find out which of my coworkers received the same question. Sometimes you'll get one customer calling several different reps because they don't like the answer they got during the last call. I usually find out about the question later in the day, but by then it's too late.

In this case the problem was impossible to solve with the existing system since calls were not logged on-line during the day but at night after they had been coded and keyed for retrieval. That same user, when asked how to solve the problem, responded:

> I would just write down all of the important calls for the day on a bulletin board in the customer support bullpen [All of the customer support reps at this location shared a common area called the bullpen.] so that you could just glance over to it and then go to the rep who took the call and talk about the problem with them immediately.

It would never have occurred to the user that the calls could also be available through a database, workflow system , or text retrieval system as they were received. This type of information can be invaluable to an educated interviewer, but it must first be offered by the users. Use of the blue sky method may be one way to encourage this. Be careful not to recommend a specific technology solution at this stage. That will serve to set expectations that you do not yet know are reasonable or appropriate. You will also be lured into designing systems based on the requirements of one or only a few users with whom you have had a good rapport. Users have to be educated regarding the availability and functionality of new technology. But to do so at this stage of evaluation and design is not only unnecessary but typically damaging to the process.

A more appropriate method for working with users during these interviews is the paper prototype. A paper prototype provides a visual representation of proposed changes in information flow without describing the details of the underlying technologies or new interfaces. The benefit of the System Schematic is now much more evident. With the System Schematic as a framework, the paper prototype can be presented to users within an already agreed upon model of the organization's existing information system. In the last example there may be a server dedicated to customer support that would easily support a workgroup workflow application for call tracking.

Throughout this process the System Schematic is modified on an ongoing basis with each successive set of interviews. In this way the schematic will evolve to reflect the organization's actual infrastructure and the transfer of information through the organization.

As the System Schematic evolves, certain trouble spots may begin to emerge as prime candidates for automation and streamlining. Again, it is recommended that the interviewer and developer of the schematic avoid the temptation to create a finalized representation of the schematic at this time. In fact, one of the advantages to this methodology is the ability to refine and evolve the System Schematic throughout the first two phases. At the end of phase 2 you will have a System Schematic that represents the collective view of the organization's existing infrastructure and workflow.

Phase 3: Finalizing the System Schematic and Identifying the New Infrastructure. The final phase of constructing the System Schematic consists of a review process and modification of the System Schematic to represent the required infrastructure for the workflow system.

The basic premise of the review is that no amount of interviewing will provide a complete picture of the organization's full information infrastructure. There are likely to be numerous situations where informal methods for the transfer and sharing of information have evolved that are not adequately reflected until the schematic is reviewed by the entire group.

The review of the System Schematic is a very dynamic and important part of the overall methodology. In fact the principal benefit of the System Schematic, as was stated at the beginning of this chapter, is the development of a common understanding of the organization's existing technology infrastructure among users, evaluators, and implementors. It is impossible to achieve this without thoroughly critiquing the perceived infrastructure.

The review should be conducted as a group session with members from each functional group involved with the proposed workflow application and many of the individuals interviewed during phases 1 and 2. It is inevitable that many assumptions about the way the existing system worked, or was supposed to work, will be invalid. The review will serve to identify the reality of the organization's infrastructure. The review process also double-checks the findings of the interviewer-designer and catches any false assumptions that may have been made along the way.

The result of the review is a completed System Schematic that precisely depicts the computing environment and the obstacles impeding the flow of information within that environment. Equally as important, you have established your first plateau and level of buy-in from the organization. The review process forces everyone to agree that "This is where we are." People will be much more willing to listen to where you want to take them tomorrow, when they are confident that you understand where they are today.

At this point you can begin assessing the viability of the existing infrastructure. Issues such as storage, networks, servers, desktop upgrades, user environments, and administration of distributed users can be evaluated and incorporated into the System Schematic. You should end up with a representation of the *new* infrastructure that will be used to support the EDMS application.

The final step of the System Schematic focuses on defining the pilot application within your project scope. Although you have already identified the sponsorship and the project scope at the outset in order to develop the System Schematic, you have also defined numerous applications and workflow cycles within this project. You must choose among these for your first implementation. The pilot should be a concrete, well-defined, and confined application that is in alignment with the CSFs. The pilot will be used to begin the organization's acclimation to EDMS. The System Schematic alone, however, does not provide sufficient information to select the best candidate application. A further exploration of the business cycles and the potential applications must be conducted before defining the pilot and the actual technology approach to be used. This will be discussed further in the next chapter on the Stair Step method.

The Stair Step Method

Because it is probable that any EDMS application will span a large portion of the enterprise's information systems, the issue of where to begin is inevitable. Traditional approaches to systems analysis and design have always employed two distinct approaches to answer this question: enterprise modeling and prototyping. These two basic approaches take many forms. For example, top-down design and programming are a version of enterprise modeling, where problems are solved from a broad framework to a more detailed one. Bottom-up design and programming use the alternative prototype method, where individual solutions are tested and implemented before moving on to a broader framework. There are problems with both of these approaches if one is used to the complete exclusion of the other.

In the case of EDMS applications, enterprise modeling not only requires a substantial analysis, due to the innate complexity of these applications, but also compromises the highly individualized nature of each workgroup's workflow and document management needs. The complexity of the undertaking is compounded by the fact that underlying office automation technologies are changing rapidly. By definition, any enterprise system implemented in one fell swoop is outdated by the time it is defined.

On the other hand, prototyping can result in a number of applications, each for a very specific set of needs, but with little consideration for communication and sharing of documents from one application to another. That is, in fact, how we created the problems described in the first two chapters of this book!

The Stair Step method of system design combines these approaches, stressing a long-term commitment to the enterprise but delivering short-term results within a limited and manageable framework. The method takes a top-down approach to system design and a bottom-up approach to implementation. The Stair Step method begins with a global view of the problem and the solution, but implements the solution one step at a time.

There is no standard definition of the size of a step. Each application or step is a process that is defined well enough so that it can be quickly and easily implemented. In each situation a step can be defined differently. In one case it could be a department-wide system and in another it could be one workgroup within a departmental system. The same methodology for defining the organization through the identification of sponsorship, recommended in the definition of the System Schematic, is appropriate for determining steps as well.

After the organizational view of the problem and the solution are determined, the individual functions that make up the processes can be investigated so that a solution system can be designed and implemented for each of the functions. The creation of further levels of detail, or steps, continues until tactical applications that can be easily worked with are defined. The actual implementation occurs at this tactical level. The design and implementation process repeats itself for each additional tactical step until the project scope identified in the System Schematic has been addressed. The design and implementation becomes easier with each new step, and the comfort level of the users and management grows with each successful implementation.

This approach may appear contrary to the popular belief in reengineering organizational systems at-large in order to achieve maximum benefits. It is not. Instead of attempting to reverse-engineer and rebuild the enterprise from ground zero, which may not be practical given the competitive pressures and ongoing business requirements of the organization, Stair Step provides an enterprise learning curve that builds on previous success and experience. The end result is an enterprise system that addresses the impossible task of predicting nuances of individual workgroups and applications, while meeting the evolving needs of the entire enterprise.

This method is very successful with EDMS because it encourages early appreciation for the impact and benefits that the new technologies will have. These benefits are often difficult to measure and appreciate until an application is actually implemented.

Moreover, the Stair Step method sets the stage for ongoing modification and fine-tuning of the system. Because the method tackles one manageable piece of the system at a time, modifications are easier to make. The Stair Step method also provides a source of information about the ways in which the system will be used at each step. This will prove to be a valuable knowledge base upon which to build each iteration of future applications.

Developing a Stair Step Model. Developing a Stair Step model begins with the creation of a matrix of relationships among the applications and document groups identified in the System Schematic. The matrix lists each workgroup along one axis and the office automation, document management, and other potential workflow applications along the other. Stair Step uses a matrix instead of the classic hierarchical model of the organization in order to avoid the trap of modeling applications after existing bottlenecks in the organizational structure. The problem with simplistic hierarchical views of the enterprise is that they obscure the complexity of information transfer, which usually circumvents organizational boundaries and structures. When associations are drawn to and from each node of a classic organizational hierarchy to demonstrate the actual transfer of information, the hierarchy quickly turns into an intricate network of relationships. Defining the individual steps of a workflow or document management process involves more than just identifying each node in the hierarchy. It requires isolating related workgroups by identifying several closely related applications or information-sharing requirements. Because of the hierarchy's underlying complexity the use of a matrix simplifies the process of grouping these applications together and makes obvious the true nature of the organization's requirements for sharing information.

Since the matrix lists organizational groups along one axis and workflow applications along the second axis, we can correlate one category to another. As you do this patterns of clustered groups emerge. Each of these clusters identifies a candidate workflow application for further evaluation. Applications which may have seemed discrete prior to the grouping are now combined into functional areas or *steps*. It should be pointed out that the term *step* in this context represents a series of applications that are implemented based on an established priority. Determining the priorities of the clustered applications and workgroups is the next phase of the Stair Step method.

The priorities are based on the criteria of payback, complexity, and visibility. A high-priority application provides the fastest payback, the lowest complexity, and moderate visibility. Although it is rare to find an application that satisfies all of these criteria, some applications will clearly come closer than others.

The first criterion, fast payback, is a function of quantifiable benefits and ease of implementation. Because of the importance of fast results in generating momentum within the organization, applications that can quickly demonstrate even moderate payback are superior to those that promise high payback only after a long implementation. A period of six months to one year is usually considered a reasonable amount of time to satisfy this criterion. Anything beyond one year increases the risk of lost user and organizational momentum. It also increases the probability of near-obsolete technology by the time it has been implemented. The ideal pilot application will create momentum through its quick demonstration of benefits and payback. Keep in mind that we are not suggesting compromising long-term success and payback. Just the opposite is true. By ensuring short-term success we will develop leverage, an understanding of the way the EDMS technologies benefit the organization, and user acceptance that allows us to achieve the maximum long-term payback. You must also consider that the complexity of most enterprise-wide EDMS applications, the lack of standards in many of these technologies, and the various levels of maturity among these technologies present a high level of risk for applications that promise only long-term payback.

The second criterion, complexity, is closely related to payback. Complexity, however, also refers to the interrelationship of applications. A candidate application with many tentacles to other applications or user organizations can prolong implementation and make measuring its success difficult since it can not be confined to a limited set of functions and expectations. The ideal pilot application will be limited in scope to provide for easily measurable success. In many cases you may want to go even further by establishing minimal expectations on the part of users and management. Your attitude should be to attack a problem that you know you can solve. Even seemingly trivial EDMS problems will be a valuable training ground for you and the organization. In addition, situations will arise and new areas of opportunity will present themselves during the pilot that will have significant value during future development efforts.

The final criterion, application visibility, helps to establish viability and helps attract useful feedback on the functionality of the application. Extremely high visibility, however, can foster excessive and unrealistic expectations. A balance must be found between these two extremes in order for the first application to act as a foundation for further development. The ideal situation calls for a moderate level of visibility that provides flexibility in developing the application and recognition of the application's value.

If you adhere to these basic criteria you will be able to quickly and objectively identify the applications that are of greatest value at this initial stage. You will also have laid the groundwork for a long-term strategy comprising a series of applications that can benefit from EDMS, resulting in an EDMS-enabled enterprise.

Once you have identified your pilot application, you can begin to set expectations for the implementation of the system and begin the evaluation of candidate product solutions. This is done by identifying the functionality required of the system in the selected applications.

The Document Audit

This stage of analysis focuses on the documents that are used in the targeted applications. The goal at this point is to determine the applicable role of imaging and/or text retrieval in your solution. This is accomplished by auditing the business documents for content and projecting this onto the potential business value of the document. The data and users' insights into the data's value must be examined. The presence of images (handwriting, graphics, pictures, etc.) and text in a document does not necessarily indicate that both technologies are required to adequately address the document management application.

Despite the presence of graphics in the original documents, you must determine whether support for the graphics is necessary. For example, the personnel department of an East Coast petrochemical company decided to manage and publish its benefits manuals on-line.

A preliminary examination of the existing paper-based manuals indicated that images and text were present. For this reason, the company assumed their solution was an integrated text retrieval and imaging-based system. A closer examination of the manner in which the documents' graphics were used by the authors and users of the documents, however, uncovered the fact that the graphics were superfluous and did not provide additional information. Though the use of graphics made the paper-based document more attractive, users never relied on the graphics for information nor did they retrieve information based on a graphic. In this case, a single technology solution (text retrieval) was used to bring the manuals on-line. Though the application offered no support for images, it completely met the users' needs for easier and faster access to information. The solution that was implemented was far less costly and more easily maintained than the integrated solution that was originally envisioned.

Similarly, documents that contain textual elements, in their entirety or otherwise, do not necessarily require text retrieval in order to provide added benefit to the organization. For example, another of my clients, a major financial institution headquartered in the Midwest, automated its receivables management system by bringing all related documents on-line. Despite the fact that many of these documents contained substantial amounts of text, the application was based on a stand-alone imaging system. It was determined that the only manner by which a document, or group of documents, would be identified and retrieved was by a known invoice number or client code. Furthermore, it was the document itself, complete with signatures and marginalia, that was of value to the user, not the textual information per se.

Even in cases where the EDMS must provide imaging as well as text retrieval functionality, product selection is not straightforward. You must determine which approach to integration is warranted by the application. Consider a litigation application in which content-based retrieval is crucial to the research process, but ultimately what is needed is a document image. This application would be best served by an imaging system with added text retrieval capabilities. On the other hand, if the majority of document management, manipulation, and retrieval focuses on the textual portion of a document despite occasional pictures and illustrations, then the product selection should focus on text retrieval products with support for imaging. The differences in these product approaches is sometimes subtle, but nonetheless important. In the first example, a more complete imaging offering is required. In the latter case, the product focus is on text retrieval. Images are linked to text documents using a hypertext-like facility and viewed in windows. The viewing of images is user-discretionary.

These products typically offer the ability to link to existing images, but provide little direct support for imaging functionality.

In either case, the need for text retrieval will warrant a second analysis focused on the nature of the research, user familiarity with the documents' contents, and the dynamics of the document database. This is necessary in order to ascertain the appropriate mix of search methodologies and query-enhancing tools to support the business application. (For more information on search methodologies, query-enhancing tools, and their relationships to user requirements, see Chap. 5.) When this is complete, the focus of the analysis turns to the business cycle itself and the potential role of workflow automation.

Collapsing the Business Cycle

Organizations are made up of a series of intricately intertwined business cycles. When considering how to streamline an organization through the use of workflow, the first things to consider are these business cycles. The objective of the workflow component of EDMS analysis is to redefine and then reconstruct the components of lengthy business cycles in such a way that the time required to execute a task is minimized and the transfer time between tasks is eliminated entirely.

Task time is addressed through reengineering itself as well as imaging, text retrieval, and workflow. Transfer time is strictly the domain of workflow. In this section we will discuss how analysis specific to workflow is executed and the criticality of it to the overall EDMS success.

For example, a major West Coast banking institution needed to streamline a policies and procedures application. Upon examination of the existing system it became apparent that new policies and procedures sometimes required over one month to be approved. When the tasks that made up the approval process were listed, however, they totaled only five days! The business cycles, on the other hand, averaged 15 days. The discrepancy was in the timing of the monthly board of directors' meetings held on the second Tuesday of each month. At these meetings every new policy and procedure had to be read into the minutes and approved by the board. The result was a relatively efficient process bogged down by the politics of the organization. This is not atypical; business cycles are often associated with organizational habits, culture, and politics. The key is to uncover these bottlenecks and redesign the process in order to overcome organizational obstacles. The application of workflow technology is a secondary issue which may in some cases not even be required once the business cycle is analyzed and redefined. It is important to understand and to appreciate this when assessing the organization's workflow and recommending technologies to help streamline it.

Almost all of the analytical tools used for traditional systems analysis lack the one critical component that must be considered in evaluating the efficiency of information-based tasks: time. Correct workflow analysis uses time-based methods in order to identify problematic business cycles. But there are other benefits to Time-Based Analysis that are less obvious but just as valuable.

First, Time-Based Workflow Analysis focuses on transfer time, that portion of the business cycle between people. This in and of itself minimizes any perceived threat of pinpointing the employee as the problem. Time-Based Workflow Analysis looks at the system and the inherent problems with the process, not at the people. This is not to say that human factors are not a key aspect of workflow. It may be that further exploration will demonstrate that there are personnel issues that must be considered, but starting with the premise that people are the problem will alienate both good and poor workers, and inhibit the analyst's ability to take an objective and complete view of the current state of affairs. Often, prolonged task times result from deficiencies in the current system's ability to transfer information in a manner optimized for peaks and lows in current user workloads. Organizations may look upon these as management-related problems that can be solved by increasing the efficiency of the task and worker. That often ignores the real problem. In many cases the elimination of transfer time can substantially improve a business cycle. Add to this workflow's ability to track the document and spontaneously interact with individuals throughout the workflow process, and suddenly it becomes possible for workflow to change the business cycle time without changing a single task, since idle times between tasks are collapsed.

Second, Time-Based Workflow Analysis provides an objective measure of the existing problems. Although it is possible to debate the inefficiencies of work habits, information systems, and infrastructure, the element of time can be objectively measured. In analyzing a process for potential workflow improvements, it is important to present objective and quantitative metrics rather than rely on intangible benefits that cannot be readily justified. This is true even when justifying workflow based on future opportunity.

Third, the process of defining task times, transfer times, and total business cycle times will expose the full breadth of inconsistency and complexity of the business cycles in question. It is not unusual to find that some business cycles have no consistent time duration. In the policies and procedures example given earlier, the time required for the approval of a given policy varied widely from two weeks to six months. What appeared to be one business cycle actually consisted of multiple formal and informal cycles. Until Time-Based Workflow

Analysis was applied it was impossible to identify the problems with the existing process or the applications for improved workflow.

Using Time-Based Workflow Analysis

Although Delphi's formal methodology and tools for Time-Based Workflow Analysis require some skill and experience, the premise and concepts can be easily grasped and applied to all aspects of a workflow analysis. The first step of Time-Based Workflow Analysis is the definition of the tasks involved in a given business cycle. The definition of the business cycle should be broad enough to include all of the activities for a certain logically connected set of tasks.

For example, publishing a policy minimally requires several tasks: initiating the new policy, reviewing and authoring the policy, executive approval of the policy, checking consistency with existing policies, board of directors' approval, typesetting, publishing, and distributing. Each task is then associated with a task time. This is the total amount of time it takes to complete the task, exclusive of the time spent waiting for information related to the task or subsequently transmitting information to the next task. For instance, the task time for board of directors' approval of a policy is never more than two hours, since this is the length of the directors' meeting. It may, however, take two weeks to actually approve the policy since the directors only meet twice each month. This is represented in the second variable, which is the transfer time between tasks.

Transfer time identifies the time required to transfer information from one task to another. At first this may seem a trivial distinction. After all, isn't transfer time part of the task? In the case of the directors' meeting, once the author writes up the policy it's no longer his or her job to get it approved; it's the responsibility of the directors. As far as the directors are concerned, however, they fulfill their responsibility through meeting twice each month. The problem becomes painfully obvious—transfer time belongs to no one. No one takes responsibility for it and no one is to blame. On the other hand, if we arbitrarily say that it is the problem of the directors to meet more often we are suddenly putting someone in the position to defend a piece of the process which they cannot control. It's no surprise that business cycles are so difficult to collapse if there is no accountability for one of the most time-consuming elements of the business cycle.

It is for this reason that we separate the two components of workflow analysis. Tasks clearly belong to individuals and workgroups. Transfer time, however, belongs to the process itself and the information systems in place to facilitate the process. Once we have established that transfer time is a process issue and not a people issue it becomes much easier to identify problems with the existing process.

By evaluating the relationship of transfer time to task time we can begin to formulate opinions as to likely workflow candidates for streamlined business processes. For example, if the total transfer time is significantly higher than the task times there is a great deal of opportunity to reduce information transfer time without necessarily changing tasks. The benefit is, of course, less disruption on the part of users. It may also be the case that transfer times vary dramatically from one set of tasks to another. This may be indicative of a bottleneck situation that reoccurs in the same area regardless of the information being processed. That may represent a point solution for automation that again minimizes user disruption.

In every case the time-based model can deliver a concise assessment of the existing workflow environment and thereby make manifest the potential areas of opportunity for a new workflow system.

Keep in mind that throughout our discussion of Time-Based Workflow Analysis we never said that people may not be the problem at some points along the process. That may well be the case and it may require dealing with as part of the workflow analysis. Starting with that premise, however, is very dangerous and time-consuming. The two-component approach presented in this chapter shifts the focus to redefining the process model. Users are much more likely to cooperate in that case and the underlying inefficiencies of the business process are much more likely to be resolved. Applying Time-Based Workflow Analysis will help you to demonstrate short-term payback that will provide long-term leverage for your workflow applications.

8 Developing an EDMS RFP

Soliciting Solutions

The quest for the Holy Grail can be found in every facet of life. In information management the search for a perfect solution comes most often in the form of the *Request for Proposal* (RFP)—that elusive document that most dread to even think of much less actually develop. RFPs are as much a cry for help as they are a search for a solution. Written by committees that are unlikely to be able to agree on what to have for lunch much less which technology vendor to choose, the RFP ends up either as the narrow description of a single vendor's technology or a "bored to tears" stream of consciousness. Both of these extremes defeat the goal of the RFP, which is to clearly define the objectives of the solution and leave some room for interpretation of technologies that suit those needs. This is a difficult but not impossible balance to strike.

The term RFP is only one of several acronyms used to identify the process of searching for a technology solution. The three most popular of these are RFI, RFQ, and RFP.

The Request for Information

The Request for Information (RFI) is intended to help establish the knowledge base of the evaluator for further investigation. Simply stated, it is a request for an education. This works particularly well in the case of technology which is established, but with which you are not adequately familiar. It will not work with incipient technologies that are still maturing. There are too many alternatives and too few standards to govern the solution. As a result each solution may appear to be justifiable in its approach. Discerning real solutions from "vaporware" becomes especially difficult in this case.

The Request for Quotation

The Request for Quotation (RFQ) assumes a solid knowledge base. Here the objective is to uncover the most cost-effective solution, with a focus on cost. RFQs are properly used in situations where there is an established technology market. There is little variance in the way of the technology or solutions that you are evaluating, so vendors will differentiate themselves primarily by cost.

The Request for Proposal

The RFP is the most complex of the three methods due to its dual emphasis on the problem statement and the solution requirements. An RFP may be used for mature and incipient technologies. Its objective is to determine how a variety of alternatives compare with each other against a list of functional and financial requirements.

Writing the RFP

Although many RFPs focus on technology, most in painful detail, the process of developing an effective RFP must begin with the business issues that are being addressed. In fact, the best RFPs will clearly define the business needs of the organization and the information technology strategy before describing the desired solution. This approach establishes the context for the correct solution, but alone, it serves only as the introduction of a well-designed RFP. After describing the business environment, the RFP must define the parameters of the solution. To many, this seems to be a contradiction in terms. An RFP does not define a solution; after all it is written in search of a solution. This is not the case, however, with technologies that are still being defined through emerging standards and market acceptance—technologies such as imaging, text retrieval, multimedia, and workflow.

There are basically two schools of thought on the development of an RFP; one says gather as much information about each vendor as possible by asking every question imaginable, and the other is to define a specific set of needs and requirements that will distinguish vendors in your context. In the case of EDMS, the second alternative is recommended. The reason is simpie. The EDMS market is still relatively immature. As a result many vendors provide apparently similar solutions, until they are scrutinized within the context of a particular application. It is then that the advantages of one solution over the others will emerge. By developing a tightly defined RFP you take firm responsibility for the definition of your solution, and thus significantly minimize the risk of inflated expectations and partial solutions. (See Fig. 8-1.)

All too often, however, the RFP is written in hopes of an answer. The philosophy of such authors is that the description of the problem in belabored detail, including data dictionary definitions, interface layouts, workflow charts, hardware listings, and user requirements will provide vendors with the information necessary for the vendor to offer a suitable solution. That is not only wishful thinking, it is a naive attempt to shirk the responsibility of analyzing the problem and researching the available solutions. Using the RFP to define a problem may seem to be easier at the outset, but it will make the evaluation of an EDMS particularly difficult. Although defining a solution requires more up-front work it is well worth the effort. To maintain that focus keep the following guidelines in mind when creating your RFP:

- Think from the top down.
- Keep it focused.
- Simplify and clarify.
- Differentiate.
- Describe the infrastructure.
- Evaluate through example

Think from the Top Down

A good RFP starts with a solid foundation, and proceeds to increasing levels of detail. In general, the format should correspond to the following levels of detail:

- Critical success factors
- Technology strategy
- Tactical applications
- Required functionality
- Alternatives
- Success metrics

By beginning with a brief mission statement rather than a problem statement, you provide the reader with a strong sense of priority and

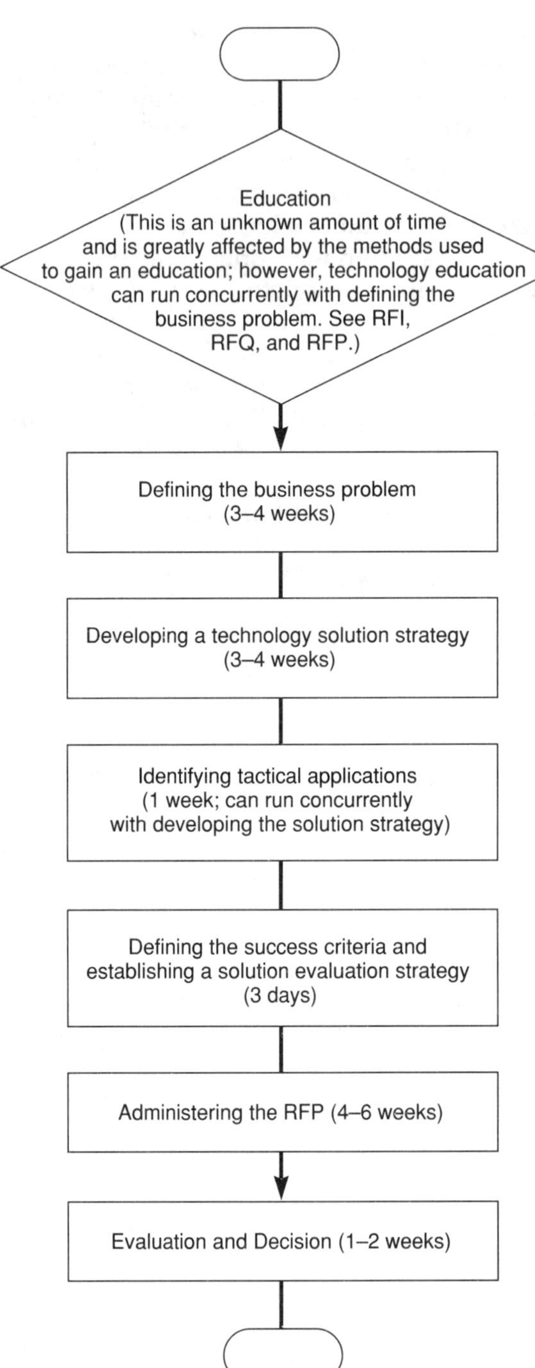

Figure 8-1 Steps to developing an RFP.

focus. This will serve as a mark of credibility for your RFP. Vendors will realize that this is a long-term solution, one which has the potential for a long-term partnership. In many cases this is important to establish since your initial application may be a small pilot that lacks the scale necessary for any significant leverage on your part.

Keep It Focused

The RFP, similar to the application it is being written for, should be well focused to avoid excessive expectations. The single most important thing you must do in writing the RFP is to define the first tactical application for the technology. This is the linchpin of your success. If there are functional requirements that support this tactical application, they should be emphasized and described as "hard stops" for responding vendors. In other words, if the vendor cannot, at minimum, meet the hard stops, there is little to no chance of the vendor winning the contract. Specifying the application will qualify the vendor responses from the outset.

Simplify and Clarify

One of the most often ignored issues in developing an RFP is that of clarity. Although you may not be especially interested in making the vendor's life easier, it will ultimately be in your best interest to be clear and concise. By using redundancy moderately, keeping the introduction terse, and utilizing ample structure, this can be achieved. For example, organize the RFP into numbered sections, and use page breaks strategically to increase its readability and reader comprehension. The RFP has several distinct components. Each one is important and should be clearly delineated in the structure of the document. For example, use the addenda for lengthy descriptions rather than cluttering the other RFP sections with this information. A well-written RFP will, at a minimum, contain the following components:

- Instructions to bidders
- Glossary of terms
- Overview
- Functional requirements
- Vendor response survey
- System acceptance criteria
- Addenda

Differentiate

Throughout the evaluation, your primary objective as an evaluator of EDMS is to differentiate one vendor from another. This is done on both a functional and an organizational level. This is especially true in the case of a large-scale EDMS, where the organizational approach and philosophy of the vendor can be as important, if not more so, than the product. The key issue is to get a sense of the vendor's approach and priorities. The time required to implement an application may or may not be a key distinction among vendors. The specific components of implementing a solution, however, will vary from vendor to vendor. This can be a major issue in the case of systems integrators who may be drawing on similar experience and proposing identical hardware systems. The objective is to find an approach that is aligned with your organization's philosophies—one that you feel comfortable with. Encourage the vendors to elaborate when they answer these questions. This will probably be one of the only areas where you can best differentiate and compare otherwise similar vendors.

Use a point-rating system that prioritizes the key evaluation criteria contained in the RFP. The point-rating system is a crucial element of the RFP. More than any other component, it will guide the vendors' efforts and responses to your request. Therefore, you must carefully consider the values you use. In some cases, you may provide vendors with only a relative ranking of items rather than a set point score. This provides a higher degree of flexibility in ranking the responses. In such cases as public agencies soliciting through an RFP, however, points are not well advised since they can be used for protest at a later date, even though the agency's requirements may have changed since the RFP was published.

Describe the Infrastructure

The proposed solution will be intimately dependent on your organization's infrastructure: hardware, software, and networks. Use the System Schematic as a tool for describing the existing and proposed technology infrastructure. Typically, the System Schematic is placed in the addenda with a narrative description of the computer environment. It may, however, be advisable to use the System Schematic within the body of the RFP as an introduction to some of the requirements or obstacles that you have identified. The approach to take must be determined on a case-by-case basis, but generally, the System Schematic itself should only be inserted in the body of the RFP if support for a complicated and/or demanding infrastructure is a hard stop.

Describe the workflow of the existing and proposed system. Since workflow can be extremely difficult to describe, a simple graphic depiction often makes the description process easier and increases the reader's understanding of the scenario.

Evaluate Through Example

Provide sample scenarios of situations that will need to be addressed by the proposed solution. The sample scenarios are often weighed heavily by many evaluators because they are the closest examples of real-world application of the proposed solution to your environment. In some cases it is wise to ask vendors to respond by providing actual screen layouts of the user interfaces for a particular application.

All of this may not make developing an RFP tops on your list of most wanted projects, but it will ease the frustration and reduce the difficulty of evaluating the responses. The unfortunate reality of creating an RFP for an EDMS is that your education can either be attained before you develop the RFP, for moderate time and cost, or after you develop the RFP, at considerable time and expense, not to mention the risk of an inadequate solution. Dismissing the quest for the Holy Grail and accepting the responsibility for defining the problem and the solution is an important step in creating an effective RFP. Ultimately, taking a disciplined and well-thought-out approach will result in an easier evaluation process, a better solution, and a solid foundation for a long-term strategic plan for EDMS.

Appendix

An EDMS Primer

This appendix provides an overview of the contents of this book. It is intended to provide the reader with a lay of the land, so to speak, which can facilitate a subsequent reading of the entire text, especially for those who have little or no exposure to EDMS. You may also consider using it as a resource to educate colleagues who do not require a thorough and complete exposure to EDMS.

Managing Documents

When we refer to an Electronic Document Management System (EDMS), we are actually referencing several integrated technologies: document management, text retrieval, imaging, multimedia, and workflow. The goal of an EDMS is to enhance and preserve the value of an organization's information resources, and in doing so, optimize and streamline other business functions. One of its key benefits is the flexibility to work with a variety of information sources and formats, many of which may not have been anticipated by the developers of an application or its users.

In an EDMS, documents are stored, transmitted, and accessed online. While magnetic media are used, most EDMS also utilize optical media such as CD-ROMs, WORMs, erasable optical discs, and videodisks for the storage of electronic documents.

The basic hardware components for a system that would manage text and images are a scanner with a resolution of 200 to 400 DPI (dots per inch), a high-capacity storage system which may be either magnetic or optical, retrieval software, a laser printer with a typical 300-DPI resolution, and a graphic display terminal with at least a 640 × 480 pixel display. Generally, you will find three common architectures for these systems. They range from stand-alone systems (which often replace or enhance the paper-based systems of an organization) to host-based systems (which either use existing system resources or make enterprise-wide commitments to the EDMS) and client/server systems (which are increasingly becoming the system of choice for many organizations due to the flexibility they offer for individualized solutions and distributed processing). Host-based systems usually represent the highest-cost solutions. They also provide the security of long-term integration with other applications available on the host platform.

Methods of EDMS: How They Work

Traditional database systems store and manage numbers and characters in fields of defined lengths. An EDMS manages any type of document—correspondence, manuals, legal briefs, personnel files, product updates, market research data, etc.—and offers rapid access to its contents. Generally, these documents are produced by word processors or electronic publishing systems, or are converted to electronic form via scanners and optical character recognition (OCR). They may also incorporate information from on-line systems, electronic mail, voice mail, appointment systems, and personal notebooks.

Let's begin our discussion of document management methods by briefly reviewing the basic approaches available:

- Imaging
- DBMS
- Keywording
- Text retrieval
- Content-based retrieval
- Hypertext
- Multimedia
- Workflow

Imaging. Imaging is probably the simplest document metaphor to understand. The image alone, however, is not going to provide much more than the elimination of the paper, which is not going to provide you with a tremendous benefit. It is best to think of imaging as the mechanism by which we will capture and manipulate one form of data that is incorporated into the document, instead of viewing imaging as a document management solution.

DBMS. A database system can be used as an adjunct to imaging, acting as a pointer or index system for the images, or it can be used as a document repository through the application of binary large object (BLOB) technology. As a BLOB, the database can store any type of document object: image, text, video, or voice. The benefit of this approach is the security and administration of the database, which in many organizations is already an established function. But a BLOB does not provide the application layers required to perform the functions of retrieval and presentation of the document objects. It only provides a repository that must then be integrated with the user applications.

Keywording. Keywording was a precursor to full-text retrieval. In a keyword system words are attached to the document through an external index. These keywords must be entered manually by an individual familiar with the content and the use of the document collection. For example, a legal or medical document collection will contain many keywords that are germane to the profession and the specific subject matter of the documents. One of the benefits of a keyword system is that the words used to index the document do not have to actually appear within the document. A certain word may epitomize the ideas or concepts expressed within a document, although that word does not appear in the document. Keywording is a very limited access means, not only because of the limit imposed in retrieving only on the keywords associated with a document, but also because constant maintenance of the keyword index is required to ensure that it adequately reflects the content of the documents over time. What is deemed relevant in a document today may not be what is considered important about the document tomorrow.

Text Retrieval. Text retrieval is a vital component of EDMS. It is this technology that makes it possible to locate and use documents based on their full content, rather than on file identifiers such as a title, author, or date. Text retrieval enables users to locate documents based on conceptual queries such as "What information is available regarding the use of radiation to treat vision disorders, especially those studies which reflect reports of dizziness or exhaustion?" Because text retrieval systems do not utilize predefined filing methods (such as keywords), they encourage rapid, spontaneous, and dynamic inquiries into large document collections. Text retrieval systems search document collections based upon user-specified words or phrases, typically joined using boolean operators and logic.

Content-Based Retrieval. Content-based retrieval is an advanced form of text retrieval. Using retrieval approaches which incorporate thesauri, semantic networks, topic definitions, and various forms of artificial intelligence, concept-based retrieval systems interpret the meaning of the document and user queries, rather than focusing on exact words. Conceptual definitions are tracked via a knowledge base that can be developed internally, or in some cases purchased from a third party. Typically, concept-based systems will rank retrieved documents by relevancy. This ranking orders documents in increasing order of importance, or relevancy, to the user's query.

Hypertext. Hypertext provides a means to link documents together in a linear fashion so that users can traverse connected nodes, or chunks, of information. The linking is a manual authoring process, where each link is defined and attached to its associated information through a hypertext dialog facility. This can be one of the most painstaking aspects of hypertext since the links can be complicated, and in some cases absolutely meaningless, if they are not authored with a fair degree of forethought and an understanding of the user's expectations and anticipated use.

Multimedia. Multimedia systems expand the types of information that can be published or referenced electronically beyond the traditional data types of text and images. Multimedia systems include support for voice and video. Multimedia systems define the model for the simultaneous and synchronized display of multiple data types (text, image, voice, and video) on a single-display station.

Workflow. Workflow provides a means by which documents can be integrated with a process that traverses individual, organizational, and technology boundaries. In order for workflow to assist in the document management process it must rely on other technologies such as imaging, text retrieval, and DBMS. It is best to think of workflow as the application-independent component of an EDMS, since it should work with all documents and document applications.

Assessing the Impact of the EDMS

The acceptance of EDMS has grown rapidly for many reasons. The EDMS prevents the misfiling and loss of information that often occurs during transcription and document routing. It allows for simultaneous viewing by multiple readers. When documents such as claims forms are accessible over a network, all users reference the same, updated information every time. Electronic documents can be retrieved using intelligent and dynamic schemas not possible in paper-based systems.

With EDMS, the retrieved documents can be manipulated—that is, automated or linked to other media such as voice or graphics—without changing the original. That is functionality well beyond the capabilities of paper-based documents.

The EDMS addresses documents which vary by source (external and/or internal) and status (active or archived). Some are used to manage published (external) documents, such as journal articles, laws, and regulations. Information service firms are a growing source of published documents, such as market reports, and news services. Text retrieval is used to search these huge databases which are available via on-line services or CD-ROM.

The EDMS can also manage internally archived information, such as evidentiary documents (copies of contracts, depositions, etc.). This information is retained in its original form for future reference or as an audit trail of past transactions.

IS managers and users responding to the annual Delphi surveys on the EDMS industry reported that their systems have brought significant, measurable benefits to their organizations. For some users, the benefits manifested in shortened and more efficient business cycles. Information moves more rapidly and the organization is more consistent, flexible, and responsive. For example, a West Coast financial institution collapsed the authoring, review, and publishing of changes to policies and procedures from six months to a matter of days. In addition, the on-line system eliminated inconsistencies in the implementation of policy between offices.

Other organizations have found that the EDMS gave them the ability to anticipate and respond to the competitive pressures of their marketplace. Using an automated EDMS system, one investment banking firm was able to capitalize on early access to information to save a key client from a hostile takeover. The analyst's daily automated text scan of the Dow Jones newswire showed a key client was vulnerable to a hostile raider. The banker contacted the client and worked up a defense strategy. The fees paid to the bank for this transaction alone, in their estimates, paid for the EDMS hardware and software costs several times over. Competing investment bankers called the same company proposing similar services only a day or so after the first proposal was already presented and accepted.

EDMS technology is causing many organizations to rethink the way they collect, assemble, and deliver information. At a major U.S. insurance provider, field agents and actuaries are using the power of EDMS technology to selectively capture information from internal documents, on-line research services, and newswires for use in their daily activities. The system draws on concept-based retrieval technologies and selects only articles that meet the user's preset screening level for delivery to the user's workstation via the internal E-mail system.

Indeed, the EDMS has forced organizations to rethink they way they define the document itself. The document has become a live entity with an active responsibility to the organization. The document is no longer viewed as a static testament of fact, but as a valuable resource, and in many cases as an active participant in the business process. Documents are presented in a variety of ways in order to best meet their business objectives at the time.

In some cases the document becomes a self-managing entity, taking control not only for its life cycle, but for the business process it represents. These document objects will: route themselves to appropriate personnel based on business procedures, monitor and enforce process deadlines, take default and alternative actions, and provide management reports regarding business cycles. The concept of the virtual document, a document that exists based on content, structure, time, application, and user is not only redefining the document, but revolutionizing the ways in which we can work with information.

There's Time to Plan Wisely

The tide is clearly turning toward the increased use of electronic information. Several years ago, only the largest and most technologically innovative corporations had some form of an EDMS in place. Today, more than half of the *Fortune* 1000 corporations we surveyed use some form of EDMS.

The time to plan for an EDMS is now. These systems are offering significant competitive advantages. In a few years, they will be the equivalent of word processors, a necessary way of doing business.

Glossary

ablative: A technology for writing and reading to a WORM. The ablative approach burns a hole or pit in the disc medium during a write operation. During a read, a lower intensity laser will reflect either away from or into the disc depending upon the presence of a pit. This is the most commonly used approach by WORM manufacturers.

actor: An individual worker in a workflow system. An individual responsible for executing tasks in a business process.

ad hoc: A workflow process model that is of temporal definition or usefulness.

agent: An automated task or process in a workflow system that can be performed without human intervention.

alloy: A technology developed by SONY for writing and reading to an optical disc, in which two layers of different metals are used in the disc. During a write operation, the laser causes the metals to fuse and form an alloy, which has different reflective properties than either of the original metals. During a read, a lower-intensity laser will reflect differently depending on the presence of the alloy.

ambiguous term: A word that can have more than one definition; a homonym. The meanings of ambiguous terms can be managed in a thesaurus using the EQUALS operator.

analog: Continuous measurement or output of physical function or quantity, such as an electrical current, voltage, temperature, or sound wave.

ANSI (American National Standards Institute): A standards setting institution, which develops and publishes standards for use within the United States.

associative memory: The process of recalling or memorizing information by logically linking it to other pieces of information. Hypertext technology is based on associative memory process.

asynchronous teams: A group of individuals who can communicate through a workflow process in which they are involved, without the requirement of simultaneous human interaction.

automatic document abstracting: The use of computer software, lexical tools, and expert systems to automatically create synopses or abstracts of documents. See also **document abstract.**

bit map/bit-mapped: The format or process by which a paper document is converted into an electronic image. Image data bits or pixels are acquired, stored, or "mapped" into computer memory, and/or displayed in the exact position as in the original view, document, or scene.

BLOB (Binary Large OBject): A database data type used to store any information that can be represented as binary data. The BLOB data type is part of a database structure which provides complete DBMS functionality for the manipulation of the BLOB item.

boolean logic: A system for representing set relationships using the operators AND, OR, and NOT. The foundation of most text retrieval query languages.

broader term: A term that represents a more general definition of another term. A thesaurus operator that allows the tracking of broader terms. Using broader-term and narrower-term operators, a thesaurus can establish a definition hierarchy. See also **narrower term.**

browser: A road map of a hypertext system, often a graphical depiction of the document network, indicating available nodes and links.

bubble: A technology for writing and reading to an optical disc. The bubble approach causes the disc media to bubble when exposed to a high-intensity laser during a write. During a read, a lower-intensity laser will reflect either away from or into the disc depending upon the presence of a bubble.

button: The graphical depiction of a hypertext link; usually an icon or specially denoted text.

CALS (Continuous Acquisition and Life-cycle Support): A U.S. Department of Defense (DOD) 1985 initiative to provide a framework for the transmission and usage of information on weapon systems in electronic form. The support of specific standards by the CALS initiative has accelerated industry-wide adoption of these standards, including CCITT Group 4 and SGML. (The acronym originally stood for Computer-aided Acquisition and Logistics Support.)

CAR (computer-assisted retrieval): A generic way of referring to any automated system for tracking and retrieving documents on-line. In a CAR system, the documents do not have to be stored on-line, as is the case in a CAR microfiche system. Text retrieval and imaging are thus subsets of CAR.

card: The basic unit of information in a hypertext system; analogous to a field in a traditional database; also known as a pad, node, or chunk.

case: An instance of a workflow procedure.

CAV (constant angular velocity): A method for formatting information on optical disc. On CAV media, data is arranged in multiple concentric tracks. See also **CLV**.

CCD (charge-coupled device): A sensor which detects reflected light from a scanned document and converts the light to an electrical (analog) signal for digitization.

CCITT (Consultive Committee International Telegraph and Telephone): A group which issues standards for facsimile and electronic image compression and decompression.

CCITT Group 3: An image compression-decompression standard which utilizes one-dimensional run length encoding.

CCITT Group 4: An image compression-decompression standard which utilizes two-dimensional run length encoding.

CCL (Common Command Language): A NISO-proposed standard text retrieval query language; also known as Z39.58.

CCS (continuous composite servo): A means of tracking data on an optical disc through the use of spiral grooves on the disc's surface.

CDDI (copper distributed data interface): A high-bandwidth (100 Mbits/s) network which, using more inexpensive copper wiring, rivals fiber's bandwidth but not its integrity over a long distance.

CD-I (compact disc interactive): Consumer-oriented data format on CD media requiring a specialized player which is connected to a television. This format is backed by Philips, Sony, Matsushita, and others.

CD-R (compact disc recordable): A form of optical storage that supports the CD-ROM standards and can be played on a CD-ROM drive, but which can be written to using a CD-R drive. Similar to CD-ROM technology, data on a CD-R cannot be erased, but a CD-R can be written to in multiple sessions.

CD-RDx (compact disc read-only data exchange): A proposed standard for full-text retrieval databases. The primary focus of the proposed standard is the interoperability of CD-ROMs.

CD-ROM: See **compact disc-read only memory**.

CD-ROM XA: A superset standard of CD-ROM for mixing audio, images, and data.

centroid: The center of a retrieval sphere in a virtual storage area or document cluster space. The centroid represents the essence of the user query in a text retrieval system that utilizes document clusters.

chunk: The basic unit of information in a hypertext system; analogous to a field in a traditional database; also known as a pad, node, or card.

closure: The ability to perceive or anticipate the end of a document, path, or sequence of ideas.

clustering: A method for logically storing documents based on relationships between documents. Documents that are most relevant to one another are logically stored close together. See also **concept-based clustering.**

CLV (constant linear velocity): A method for formatting information on optical disc. On CLV media, data is arranged in one continuous spiral track. CLV has a higher storage density but slower access than CAV media. See also **CAV.**

COLD: See **computer output to laser disc.**

COM: See **computer output to microfilm.**

compact disc: A trade name for the optical read-only disc (12 cm in diameter) originally developed by Sony and Philips for audio.

composite video: A digital signal that contains all color information in one signal.

compact disc-read only memory: A version of the compact disc for storage of digital data. Its storage capacity is 556 MB.

compound document: A document that contains information in several formats—text, graphics, and image.

compound term: A multiword term or phrase. Compound terms can be treated as single terms (a word) for purposes of query submission and indexing.

computer output to laser disc: The capability to directly produce laser disc images from computer-generated signals.

computer output to microfilm: The capability to directly produce microfilmed images from computer-generated signals.

concept-based clustering: A full-text retrieval search methodology based on document clustering. Documents are logically stored in a virtual storage area (cluster) based on document content and the defined relationship of terms to subject matter. A query is placed in the storage area using the same algorithms used to position the documents. The location of the query becomes a centroid around which a retrieval sphere is formed. Documents that are within a specified distance to the query are retrieved as relevant.

concept-based retrieval: A textual search based on a concept, rather than an exact word match. A concept automatically defines a list of search terms, phrases, and rules. Accomplished at various levels of sophistication through various tools and methodologies (e.g., a synonym list, thesaurus, clustering). Also known as **concept searching.**

concept searching: See **concept-based retrieval**

control console: A visual display of the work in process including measures of productivity and throughput in a workflow system.

deengineering: A process which breaks down a business process into its individual tasks, procedures, and rules.

DIBACOS (dictionary-based compression system): A text compression algorithm based on the premise that two-thirds of the words in any document are represented by 1024 of the most commonly used words in the English language. DIBACOS can typically compress text by 60 percent.

digital paper: A form of optical storage that resembles magnetic tape, but is made of a polyester. Written to optically, a 12-in reel of digital paper can store up to 1 terabyte of information.

digitize: The process of converting a paper document to an electronic image. A document is *digitized* using a scanner.

DIMS (document image management system): See **imaging.**

DIP (document image processing): See **imaging.**

direct access storage device (DASD): A storage peripheral, usually a disk drive, which can respond directly to random requests for information.

dithering: A method used by scanners to simulate shades of gray. Geometric pixel patterns of various sizes are created according to the brightness of the image area. Also known as electronic screening.

document: A collection of information objects formatted for presentation and consumption by an end user.

document abstract: A synopsis of a document's content which highlights key concepts and significant data. The abstracting process has traditionally been performed manually. Recent advances in technology have made the automation of the process possible. See also **automatic document abstracting.**

DPI (dots per inch): A means to measure image density and resolution; for example, the number of pixels per inch on a CRT screen.

DRAW (direct read after write): Method used to ensure that data written is error-free.

DTD (document type definition): The start of an SGML document, in which the document's components and structures are defined. See also **SGML.**

duration: The time required to complete an event or task.

DVI (digital video interactive): A proprietary technology for full-motion video in real time. The DVI capacity is 72 minutes of video, with compression, on a single CD-ROM.

electronic sticky notes: Bodies of text attached to a document by hypertext links. These texts are used to annotate an electronic document without altering the document itself.

event: A task or a route.

FDDI (fiber distributed data interface): High-bandwidth (100 Mbits/s) fiber optic network. Capable of simultaneous transmission of text, image, and voice.

folder: A workflow structure that contains one or more documents related to a particular workflow task or case. A logical organization of information objects. Folders may be represented through the use of file folder, file cabinet, file drawer, and other file management metaphors.

fractal image format (FIF): A software-based image compression algorithm that maps the image into multiple pieces or areas and breaks down each area into a series of common pieces represented by a single equation and unique sets of data. This allows for the storage of only a small number of equations and data sets which represent size, shape, and color. The image is recreated by plugging the data into the equations.

free-text scanning: A full-text retrieval search methodology that "reads" the entire text database for each query submitted. Typically utilizes a text array processor (TAP) to facilitate the text scanning process.

front-key compression: A compression algorithm applied to the inverted index created by some text retrieval systems. Front-key compression eliminates the redundancy in the prefixes of the words that are stored in alphabetical order in these indexes. Compression can typically result in an inverted index of 100 percent the size of the text in the database.

full-text retrieval (FTR): A software or hardware process that retrieves textual documents based on the words, phrases, or concepts contained in the documents. Also known as *text information management systems* and *text retrieval systems.*

fuzzy logic: The application of generalized set theory logic. Fuzzy logic permits partial adherence to a criteria. In full-text retrieval, fuzzy logic often takes the form of soft boolean operators and is used in the process of relevancy ranking. See also **soft boolean operators.**

GOSIP (Government Open System Interconnect Profile): A U.S. government set of standards for wide area networks (WANs), local area networks (LANs), E-mail, documents, EDI, and FDDI.

go word: A word specifically identified for inclusion in an inverted index. The use of go words permits user definition of a key word system. Go words are the opposite of stopwords.

grayscale scanner: Scanners that can accurately sense, differentiate, and encode intermediate shades between black and white in scanned images.

group 3: See **CCITT Group 3.**

group 4: See **CCITT Group 4.**

groupware: Software that automates and manages a single task among multiple workers.

HDTV (high-definition television): A video standard noted for increased vertical and horizontal resolution, color information, and audio quality.

heuristic associations: A method for increasing the recall of a full-text query. Heuristic association is an iterative process in which the documents identified through one query are statistically analyzed to determine prevalent words and phrases not included in the original query. These words are used to initiate a subsequent query.

High Sierra: See **ISO 9660.**

Huffman coding: A generalized compression routine. Huffman codes can be generated at a word or character level. The text in a database is surveyed in order to compile statistics regarding word or character usage. A tree structure is built in which the more frequently a word or character is used, the further up the tree it is located. Bit values are then assigned to the nodes of the tree by assigning a "0" to all left branches and a "1" to all right branches. The paths to each individual word character or word become its bit representation.

Huffman tree: The tree structure built by the Huffman coding process.

hypermedia: An extension of the hypertext concept. A hypermedia system supports the linking of nontext nodes (images, graphics, audio, and video).

hyperspace: The disorientation experienced when using a poorly authored hypertext system. Usually caused by an overabundance of links and/or subjective links.

hypertext: A text retrieval search methodology based on associative memory process. Hypertext organizes textual information into sections known as nodes or chunks. The nodes can be linked together via macros that are automatically built by the system, based on user input.

HyTime: A standard hypermedia time-based language extension for SGML.

ICR (intelligent character recognition): An advanced form of OCR technology. ICR typically uses sophisticated lexical tools which increase the success rate of the data conversion process. See also **OCR.**

imaging: The process of capturing, storing, and retrieving information, regardless of its original format, using micrographics and/or optical disc technologies. See also **DIP** and **DIMS.**

index term: A word or compound term stored and controlled in an inverted index. See also **inverted index.**

initiation: An event which triggers a task.

interface standards: Industry standard hardware, software, and bus configurations; data transfer protocols; and data formats. Main standards are SCSI, parallel RS-232, and Centronics.

International Standards Organization: See ISO.

Internet: The worldwide interconnected system of networks in which all networks are reduced to a single common protocol allowing the appearance of a single contiguous network. The Internet gives the ability to share data across the world.

inverted index: An index created from an inversion of all the words contained in all of the documents in a text database, arranged in ascending alphabetical order.

ISDN (integrated services digital network): A fully digital telephone line that can transmit voice, image, and data over a single cable.

ISO (International Standards Organization): An organization, located in Geneva, Switzerland, devoted to developing the OSI Reference Model, which is a framework for standards allowing the open exchange of information among terminals, computers, networks, and applications.

ISO 9660: The CD-ROM logical file format standard adopted by ISO in 1987. ISO 9660 describes a table of contents, but not the format of the actual data. This has led to incompatibilities between different computers. Based on a specification developed by the High Sierra Group (HSG) which included Apple, Microsoft, 3M, Philips, Hitachi, and DEC, it is also known as *Yellow Book* and *High Sierra.*

JPEG (Joint Photographic Experts Group): A committee chartered by CCITT and ISO. JPEG is also the name of the standard proposed by this organization for the compression of color still images.

jukebox: An automated storage device housing multiple optical discs and one or more read-write drives.

knowledge-based: A workflow model that incorporates heuristic information about ongoing processes into the workflow process model, through the use of an inference engine, artificial intelligence, or other means.

LAN (local area network): A network of connected devices within a small area such as a single office or building.

laser: A device emitting a highly coherent beam of light for burning and reading information on optical discs.

latency: The delay in accessing data which comes from waiting for a disc to rotate to the correct location.

LD-ROM: A media format developed by Panasonic in 1991 which combines digital data on a standard videodisc.

lead term: See **use term**.

lexical analysis: A series of rules, algorithms, and tools that decompose text into its definitive parts (words and phrases).

link: The connection from one node to another in a hypertext system. Links are depicted by buttons.

link reference: The button representing the beginning of a link; the origin of a link.

link referent: The button representing the end of a link; the destination of a link.

lossless compression: Any compression algorithm that is capable of recalling all of the original data in a compressed file. CCITT 3, CCITT 4, and Huffman are examples of lossless compression.

lossy compression: Any compression algorithm which loses some of the original data during the compression of a data file. JPEG and FIF are examples of lossy compression.

microfiche: A 105 × 148 mm piece of microfilm on which images are arranged in a grid-like fashion.

microfilm: A file-grain, high-resolution film used to record images reduced in size from the original.

microform: The general term for referring to a medium that contains microimages.

MIDI (musical instrument digital interface): A hardware and software specification for exchanging information between musical instruments and related devices (sequencers, light controllers, mixers, etc.).

MPC (multimedia PC): A standard hardware configuration for multimedia. The minimum configuration consists of a 10-MHz 286 CPU, 2 MB of RAM, a CD-ROM with CD-DA output, a 30-MB hard disk, a 1.44-MB floppy disk, VGA video screen, an 8-bit audio sampling, music synthesizer, MIDI, on-board audio mixing, DOS 5.0, and Windows 3.0 with multimedia extensions.

MPEG (Moving Pictures Expert Group): An ISO committee. Also the name of the compression standard proposed by this committee. The compression standard is designed for moving images, such as video.

multimedia extension: Developed by Microsoft and IBM for DOS/Windows and OS/2, it includes a media control interface for devices such as CD-ROM and MIDI, and the resource interchange file format.

narrower term: A term that represents a more focused definition of another term. A thesaurus operator that allows the tracking of narrower terms. Using broader-term and narrower-term operators, a thesaurus can establish a definition hierarchy. See also **broader term.**

***n*-grams:** Unique strings of characters found in text. The level of granularity of an *n*-gram system determines the type of grams tracked. For example, a uni-gram system tracks each single character in the text, a duo-gram system tracks all unique pairs of characters, etc. Also known as **suffix arrays.**

NISO: An ANSI-affiliated standards group responsible for library, information services, and publishing standards.

node: The basic unit of information in a hypertext system; analogous to a field in a traditional database. Also known as a card, pad, or chunk.

noise word: See **stopword.**

nonpositional inverted index: A full-text retrieval search methodology in which an inverted index is created that tracks the presence of words in documents, but not their location in the documents. See also **positional inverted index** and **inverted index.**

normalization: The tracking of irregular extensions of root words.

notification: A workflow action that sends a message to an individual or agent. A notification may not necessarily trigger a task.

NTSC (National Television System Committee): The broadcast standard used in North America and Japan. NTSC transmits at 30 frames per second.

object: A combination of rules, procedures, and information in a single entity.

OCR: See **optical character recognition.**

ODA/ODIF: An architecture for preserving document formatting and character attributes within a compound document, making it possible to transfer documents across various computer platforms. ISO standard 8613.

omni-term skewing: A relevancy-ranking algorithm used by some text retrieval systems. In omni-term skewing, a constant value or skewing factor is added to the relevancy value of any document that contains at least one occurrence of each query term.

optical card: An optical storage media that resembles a credit card.

Developed and distributed primarily by Drexler and Canon, each optical card has a maximum storage capacity of 1.5 MB.

optical character recognition (OCR): Software technology that processes text within a scanned image and converts it into ASCII or word processing format. See also **ICR.**

optical disc: A disc with a recording alloy on which binary information is written using a laser. Various forms of optical discs exist including CD-ROM, WORM, and erasable optical.

OSI (open system interconnect): A set of interface standards being developed by the ISO.

pad: The basic unit of information in a hypertext system; analogous to a field in a traditional database. Also known as a card, node, or chunk.

PAL (phase alternate line): A broadcast video standard used in Europe. PAL transmits at 25 frames per second.

pattern recognition: A full-text retrieval search methodology in which patterns created by the groups of letters in the document text are stored and indexed as bit vectors. User queries are processed using the same pattern-creation algorithms. The retrieval process looks for similarities between the patterns in the query and in the text.

phase change: A technology for writing to an optical disc. In phase change technology, during a write, the laser causes the disc medium to either crystallize or become amorphous. During a read, a lower-intensity laser will reflect either away from or into the disc depending on its current state.

photo CD: A format being developed by Kodak especially for high-quality images, but it can also be used for data.

pixel (picture element): The smallest display element on a CRT; the more pixels the higher the resolution.

planetary camera: A digital or microfilm camera in which the document being photographed is on a plane surface, and both the document and the film remain stationary during exposure.

positional inverted index: A full-text retrieval search methodology in which an inverted index is created that tracks the exact location of words in documents. See also **nonpositional inverted index** and **inverted index.**

posting: A data record in a positional inverted index. These records track the location of each index term in every document.

precision: The measure of a text retrieval system's ability to return only those documents that are relevant to a user query. Precision is calculated as the ratio of relevant documents retrieved to the total number of documents retrieved.

predecessor: The task or route immediately prior to the current task, or route in a workflow process model.

preferred term: See **use term.**

priority: A process attribute that determines the sequencing of information objects through a workflow.

procedure: A single logically interrelated set of tasks. A subset of a process model. Procedures may be represented as single objects in some workflow systems.

process independence: The ability of a process to be independent of the underlying technologies that enable it. Process independence allows for a process to be automated across what would otherwise become technology boundaries.

process model: The highest level of definition for a workflow process. A process model consists of a sequence of tasks performed in an orchestrated manner by a variety of workers. Process models may consist of subprocesses, which are independent process models.

proximity searching: The search for one or more terms in a document within a specified distance from another term (e.g., within the same paragraph, within the same sentence).

query by example: The ability to submit an existing body of text as a full-text query.

query by form: A text retrieval query facility in which query construction is facilitated through the use of "fill in the blanks" screens and menus.

query language: A defined set of syntax and commands used to submit queries to a text retrieval system.

queue time: The elapsed time from the point when information is ready for a task to be performed against it and the point when that task begins. Queue time always follows transmit time.

queued memory: The process of recalling information through prompts or clues relating to the information. Most text retrieval search methodologies are based on the queued memory process.

raster graphics: Computer images. Digital pictures stored as a series of zeros and ones; a bit map.

recall: The measure of a text retrieval system's ability to deliver all documents relevant to a user query. Recall is calculated as the ratio of the documents retrieved to the total number of relevant documents available.

receivership: The ability to organize communications around the information requirements of the recipient instead of the sender. See also **user profiles.**

related term: A term that bears greater insight to another term, but is not synonymous or a broader or narrower definition of that term. Related terms can be tracked in a thesaurus using the related-term operator.

relevancy ranking: A feature of a text retrieval system which lists retrieved documents in descending order of the level to which they are pertinent to the query statement.

rendezvous: The point at which multiple routes from multiple tasks in a workflow process model are rejoined to proceed as a single route.

resolution route: The path an information object or a process takes in order to reach a conclusion. There are two types of resolution routes: 1 to *n*, which requires more than one task; and one and done, which requires only one task.

role: A group of one or more individuals in a workflow environment who share common responsibilities. A specific set of skills required to perform a given task.

routing: The logical, defined transfer of information through a process and its associated tasks, based on specified rules. There are five possible routing architectures in a workflow system: *serial,* in which each task has only one predecessor and only one successor; *parallel,* in which a group of tasks have the same successor and the same predecessor task; *concurrent,* which is the same as parallel, but the tasks between predecessor and successor must begin and end at the same time; *conditional,* which is a situation where multiple routes may be followed based on a rule, procedure, or variable; and *dependence,* which is a route that is predicated on the completion of another task. In effect, every route is dependent on its immediate predecessor task being completed, however a dependence routing scheme explicitly states this dependence on tasks which may not be an immediate predecessor task.

rule: A workflow procedural parameter that determines the action taken against an information object.

run length encoding: The basis for most data compression methods used in imaging. Run length encoding reduces an image's black and white regions to short codes, rather than separately storing each black or white pixel.

RVD (rewritable videodisc): A nonstandard magneto-optical disc format used by the broadcast industry. There are two competing formats developed by Pioneer and Sony.

sampled servo: A means of tracking data on an optical disc.

SAR (storage and retrieval unit): A micrographic device that houses up to 300 rolls of microfilm. Through robotics, a SAR retrieves the correct film cartridge and scans the desired image.

scanner: A hardware device that converts analog images into digital images. A scanner uses a narrow beam of light to digitize a document into a stream of bits.

scope note: Information regarding a thesaurus term. Scope notes can provide information regarding term qualities such as dated definitions, and are tracked via the scope note operator.

script: The rules and procedural definitions of a workflow.

SCSI (small computer system interface): (Pronounced *scuzzy.*) A general-purpose interface for any peripheral device that supports the interface convention such as a scanner or optical device. An SCSI chain supports up to eight logical devices, although multiple chains can be daisy-chained.

SECAM (Sequential Coleur Avec Memoire): The color television standard used in France and eastern Europe.

semantic network: A text retrieval methodology that provides concept-based query. A semantic network is a series of intersecting sets that tracks the meaning of words, their relationship to other words, and the meaning of word phrases.

SFQL (Structured Full-Text Query Language): A proposed standard for full-text databases. The primary focus of the proposed standard is interoperability of CD-ROMs. SFQL is based on the SQL (Structured Query Language) standard for relational databases.

SGML (Standard Generalized Markup Language): A metalanguage that describes the language used by a text processing package to format text. SGML preserves text format in a document when it is transferred between platforms and applications. ISO standard 8879.

single point of access (SPOA): An EDMS system configuration that supports the concept of the empowered user. In an SPOA system, each user has the ability to access any and all information accessible from the system through intuitive tools without concern to data location, data structure, or storage facility.

SMDL (Standard Music Description Language): Proposed ANSI standard extension to SGML for the representation of music in computer-based documents.

soft boolean operators: An application of fuzzy logic in which the boolean AND is treated as an OR for retrieval purposes. Soft boolean operators are used by relevancy-ranking routines; documents that are retrieved by nature of the soft operators are ranked less relevant than those that meet the exact boolean criteria.

soft hypertext: Functionality provided through the integration of a full-text retrieval engine with a hypertext system. The full-text retrieval engine allows any word or phrase in the document text to function as a button by automatically linking to all other occurrences of that word or phrase in the system.

soundex: A search based on the phonetic sound of a query term rather than its spelling or definition.

splicing: A feature of object-oriented workflow systems that permits the insertion of an ad hoc workflow route into a spliced task in a transaction-based process model.

spin-up: The time it takes a drive to accelerate a disk to its operational speed.

status: The measurement of a work in progress in a workflow environment.

stemming: A process that reduces all query terms and document terms to their common root word.

stopword: A commonly occurring or inconsequential word which is excluded from an inverted index (e.g., and, the, but, so, a). Also known as a **noise word.**

structured data: Data that is physically defined and stored in fixed-length fields. Typically, data that is managed by a traditional DBMS.

successor: The task or route immediately following the current task or route in a workflow process model.

suffix arrays: See *n*-**grams**.

suffixing: A form of wildcarding where a query term is provided as a root word followed by a wildcard character. Suffixing is automatically achieved through **stemming.**

suspense: The halting of a process or an information object prior to completion. A workflow queue which is a holding area awaiting the initiation of a task.

synonym: A word or phrase that has the same or nearly the same definition of another word or phrase.

synonym file: A file containing words grouped by similar definition. Synonym files allow text retrieval systems to search for many words that have a similar or the same meaning as the user-supplied search term. Used to increase a system's recall. See also **synonym.**

task: A finite set of actions that have a defined initiation and conclusion.

task queue: The logical storage area for an information object waiting for processing.

task time: The required time to complete a given task.

tellurium: The metal most commonly used in optical discs as a recording medium.

term: A word or phrase; usually used when referring to words in a thesaurus, synonym file, or query statement.

term density: A value used by relevancy-ranking algorithms, measured as the ratio of query terms to total number of terms in a document.

term proximity: A value used by relevancy-ranking algorithms, calculated as the distance between query terms contained in a document.

term weight: A numerical value assigned to a term representing its importance to a database or a query. Term weights can be assigned manually or automatically.

text array processor (TAP): A hardware unit that provides parallel processing specifically designed for free-text scanning.

text information management system (TIMS): See **full-text retrieval.**

text retrieval system (TRS): See **full-text retrieval.**

thesaurus: ANSI standard Z39.19; a vocabulary control facility. A compilation of words and phrases that tracks synonymous, hierarchical, and other relationships. Thesaurus terms provide a means of concept searching and control of precision and recall. See also **scope note, broader term, narrower term, ambiguous term, use term,** and **related term.**

TIFF (Tagged Image File Format): An industry standard developed by Aldus for image file storage.

transaction-based: A workflow process model that is used for repetitive processing of like transactions.

transfer time: The time that elapses between the completion of one task and the initiation of its immediate successor task. Transfer time is composed of queue time and transmit time.

transmit time: The time required to physically send an information object to its immediate successor task queue.

traverse: The enacting of a hypertext link, typically caused by enacting a link reference.

trigger: An event that initiates a task in a workflow process model.

TWAIN: A proposed standard for cross-platform scanner-to-software communications. (TWAIN is an acronym which does not stand for anything, but alludes to never the TWAIN shall meet.)

unstructured data: Data that is of variable length and subject to a change in size. Data which cannot be stored in a fixed-length field without an element of distillation.

use term: The main or preferred term in a synonym list or thesaurus entry. In a thesaurus the USE FOR operator is used to track these terms. Also known as a *preferred term* or *lead term.*

user profile: A continuously running text retrieval query submitted in background. User profiles will typically notify their respective owners of any new submissions to the text database that satisfy the query.

value chain: A series of tasks or processes which add value to an information object as it progresses from task to task. A value chain can be represented by organizational groupings as well as process groupings.

vector graphics: The technique of digitizing a CRT image by using only points (vectors) of the image. A vector graphics square would contain four sets of coordinates. Vector graphics are preferred for computer-assisted design (CAD).

VGA (Vector Graphics Adapter): An industry color graphics display standard. Standard VGA is 640 × 480 pixels.

videodisc: An optical read-only disc that stores 30 to 60 minutes of audio and video in analog form.

WAIS (Wide Area Information Server): A commercially available implementation of the Z39.50 ANSI standard. WAIS expands the functionality of Z39.50. Its source code is available over the Internet. WAIS controls the transfer of information of any type across a client/server network.

wildcarding: The use of characters known as wildcards in a query term which signify that any group of characters can be used to complete the term.

workflow: A proactive toolset for the analysis, compression, and automation of business processes.

work queue: A holding area for information objects while they await processing or are being processed in a workflow.

WORM (write once read many): A characteristic of any digital storage medium on which information can be recorded once and read many times thereafter.

Yellow Book: See **ISO 9660.**

Z39.50: An ANSI-NISO standard query language that controls the retrieval of information (including text) in an interconnected environment.

Z39.58: A NISO-proposed standard text retrieval query language. See also **CCL.**

Index

Note: The *f.* after a page number refers to a figure; the *n.* to a note; and the *t.* to a table.

Thomas Koulopoulos

Carl Frappaolo

About the Authors

THOMAS M. KOULOPOULOS president and cofounder of Delphi Consulting Group, is an internationally known lecturer and consultant on Electronic Document Management and Workflow, as well as developer of the Time-Based Analysis method. Recognized worldwide as the industry authority on the subject of workflow and technologies for business process design, Mr. Koulopoulos was named one of the top six consultants in the information industry by *InformationWeek* Magazine.

CARL FRAPPAOLO executive vice-president and cofounder of Delphi Consulting Group, is the industry's leading authority on the technological and practical aspects of text retrieval and document management systems and methodologies, EDMS integration, and large-scale applications. Frappaolo is considered a visionary on the nature of the electronic document and the role it plays in redefining the office. As Delphi's principal consultant, Frappaolo has designed solutions for a wide variety of organizations spanning multiple industries.